CARS

Table of Contents

Easy Ideas 1
Airplanes 2
Cars 3
Computers 4
Smartphones 5
Food 6
Nature 7
Space 8
Light 9
AI 10

STEM-Zen Program

© Copyright
Indē Ed Project
Non-Profit, Charitable Org'n
2023. All rights reserved.

Everyday Objects
Bonus

Easy Ideas From Concepts to Critical Thinking 1	**Airplanes** From Four Forces to Flights 2	**Cars** From Actions to Autos 3
Computers From Digital to Data 4	**Smartphones** From Calls to Global Connects 5	**Food** From Eats to Energies 6
Nature From Atoms to All Life 7	**Space** From Elements to Us 8	**Light** From Suns to Sapiens 9
AI From Machine Muscles to Minds 10	**STEM-Zen Program** From Empty to Science EnLights	**Everyday Objects** From Ideas to Daily Items Bonus

Cars — Four Simple Points

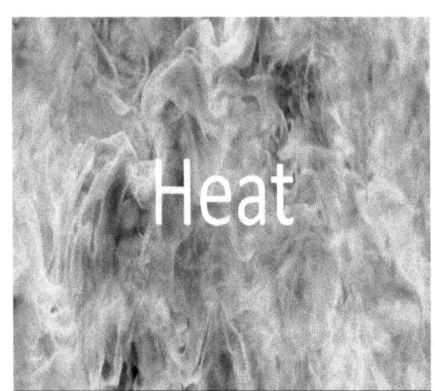

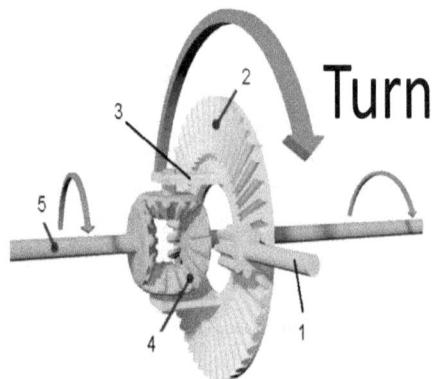

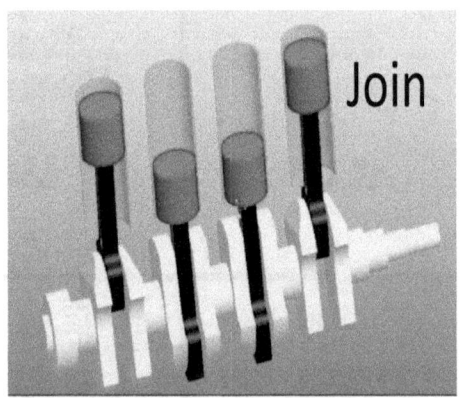

Put Pieces Together

How Cars Work — Burn and Turn

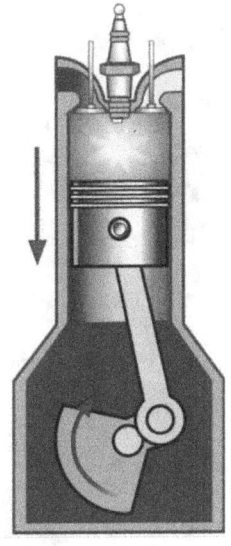

Gas burns and gives off heat and pressure. Fire bursts push the pistons down. The pistons turn the crankshaft. Gears and shafts transmit turning to the tires.

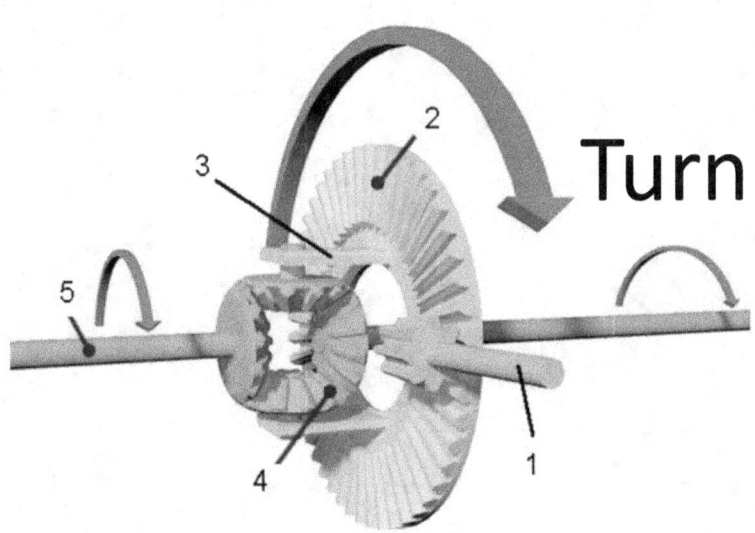

Turn

3) Cars

−Teacher Guide−

Science

3) Cars

Science pours, pounds and pushes metal and plastic into car parts. Cars have to be made before they can move us around!

  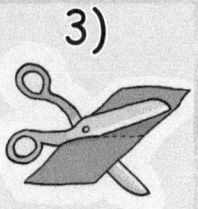

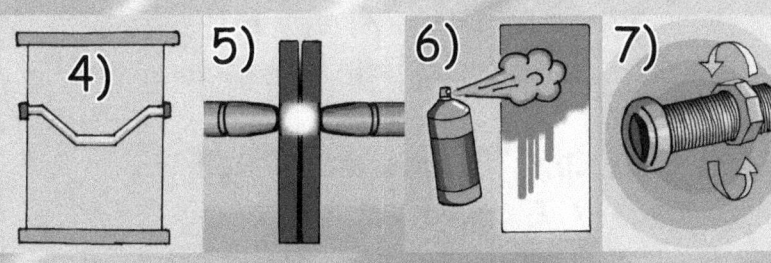

Table of Contents

Factory
Power
Electricity

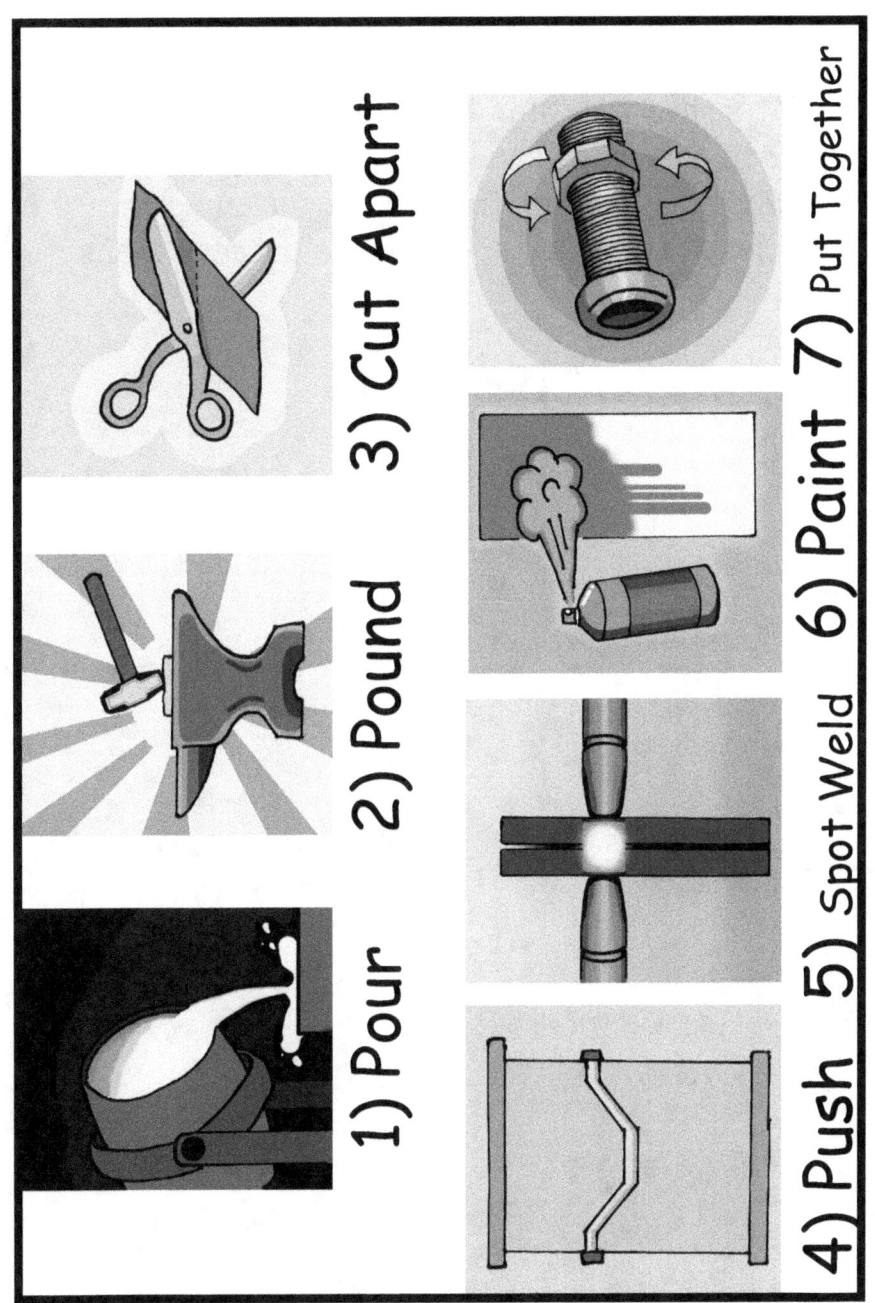

1) Pour   2) Pound   3) Cut Apart

4) Push   5) Spot Weld   6) Paint   7) Put Together

B) <u>Roadmap</u> — Cars

Purpose: Science enables car parts that integrate together to energize automobiles. Each part adds important capability to the whole function of the car. Focus on how science is applied to the make process steps.

Main Points

0) Heat and pressure
are important to how cars are made.

1) Heat melts the metal that
is <u>poured</u> into molds to make engines.

2) Pressure <u>pounds</u> hot
metal to make crankshafts.

3) Wedge-shaped tools
<u>cut apart</u> gears from room temp metal.

4) Pressure <u>pushes</u> sheets of flat, thin metal between tools to make body frame parts.

B) <u>Roadmap</u> — Cars

5) Electricity heats spots in frame pieces to <u>spot weld</u> them together.

6) Negatively charged <u>paint</u> is attracted to the positively charged frame assembly.

7) All the pieces are <u>put together</u> or assembled to complete the car.

8) Cars go because gas burns in the engine. This causes heat and pressure. As the gas burns it puts pressure on pistons that turn the crankshaft. Gears transmit the turning to the tires.

B) <u>Roadmap</u> — Car Parts

Cars are made of parts.

1) Parts are made with science
   like heat and pressure.

2) <u>Heat</u> is energy that
   flows from hot to cold.

3) <u>Pressure</u> is when force
   pushes on something.

4) Parts (ingredients) are
   <u>put together</u> to make things
   like cookies, clothes and cars.

C) <u>Refresh</u> KEY CONCEPTS — Cars

1) Pour — Cast

Pour or cast, very hot liquid metal into a mold.

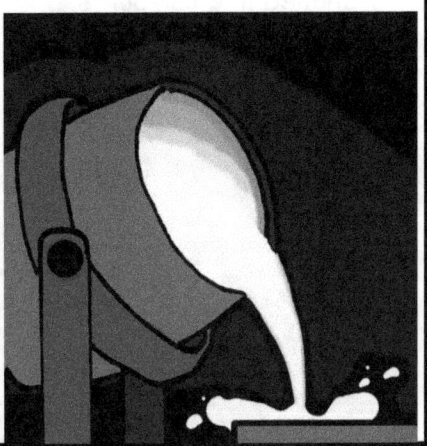

2) Pound — Forge

Pound or hammer to forge hot metal into the shape.

3) Cut Apart — Chips

Gears change or transmit turning. Gears are circles with teeth. Tools cut apart chips to make gear teeth. This is similar to scissors cutting paper or a knife peeling an apple.

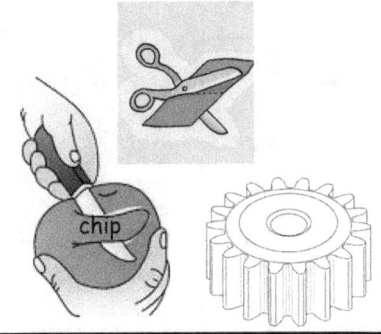

C) <u>Refresh</u> KEY CONCEPTS — Cars

4) Push Form—Frame Parts

Form is to push sheet metal pieces into tools shaped like car parts.

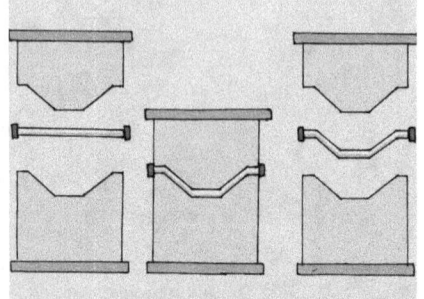

5) Spot Weld

Spot weld uses electricity to melt points in metal parts. The liquid hot spots, cool and join the frame parts.

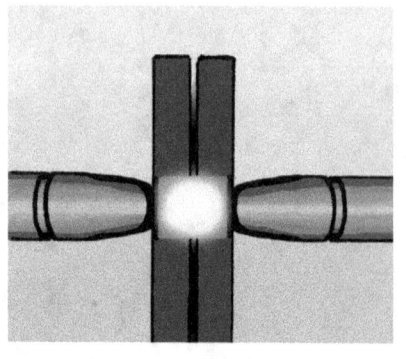

6) Paint

Paint the car in three steps: 1) Under, 2) Middle and 3) Top Coats.

7) Put Together

Next, put together or assemble the car using a balance of humans and machines. This automated assembly tool installs the major systems.

See the eBook on how to make cars in 7 steps.

C) <u>Refresh</u> KEY CONCEPTS — Cars

Processes

Processes are actions that make things like cars.

A) Pour
B) Pound
C) Cut Apart
D) Push
E) Spot Weld
F) Paint
G) Put Together

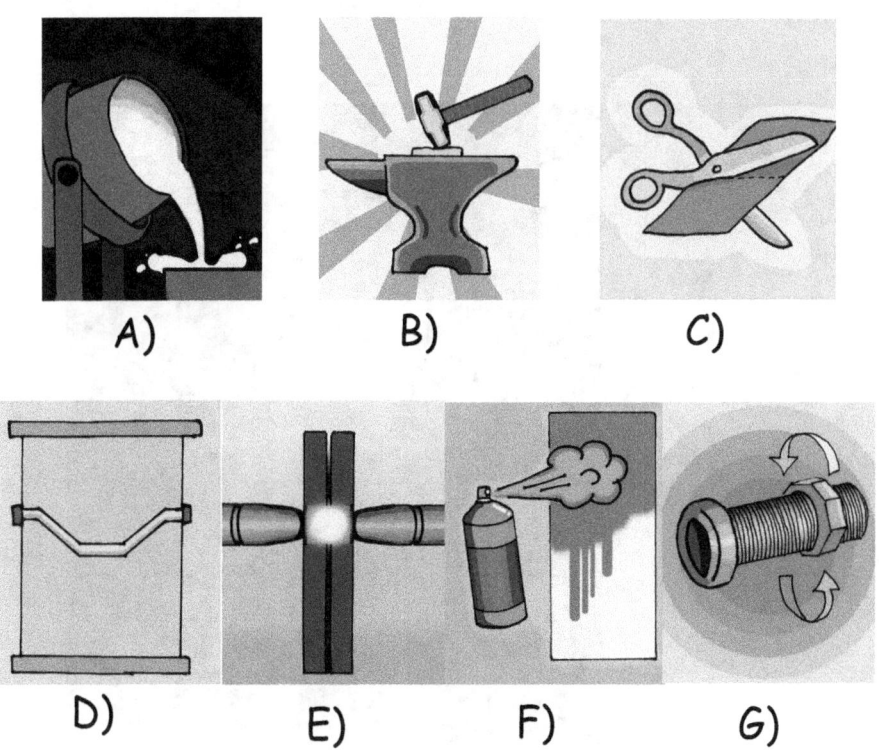

A)   B)   C)

D)   E)   F)   G)

Science enables and underpins processes.

HEAT

Heat melts ice into liquid water.

Heat melts metal into a liquid too.

Teacher - Cars

PRESSURE

Pressure applies force to objects.

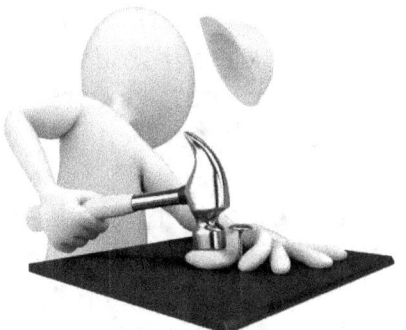

Hammer nails.

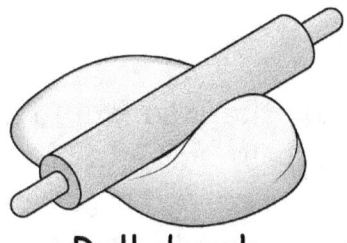

Roll dough.

Chop veggies.

Cars Have Science Inside

Explain to the students that
with cookies and clothes we see
how atoms change into parts.
Then, parts are put together to make
our everyday objects called products.

Next, we are going to start with the
finished car. We go backwards
through the processes or sets of
actions that make cars.

Encourage the students to notice
how different amounts of <u>heat</u>
(temperatures) and <u>pressure</u>
(forces) are used to make cars.

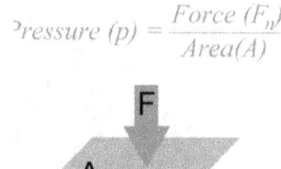

$$Pressure\ (p) = \frac{Force\ (F_n)}{Area(A)}$$

To review, action steps that make things are called processes.

Heat and Pressure

Matter at different temperatures and pressures can be shaped into car parts.

Heat   Pressure

Engine Energy

Cars go because of engines. Let's pause and see how engines work.

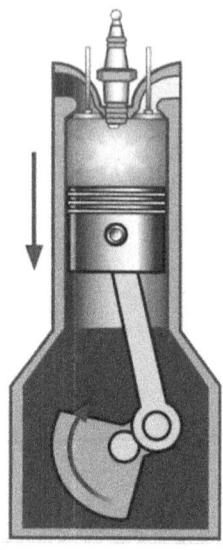

Gas burns and gives off heat and pressure. Fire bursts push the pistons down. The pistons turn the crankshaft. Gears transmit turning to the tires.

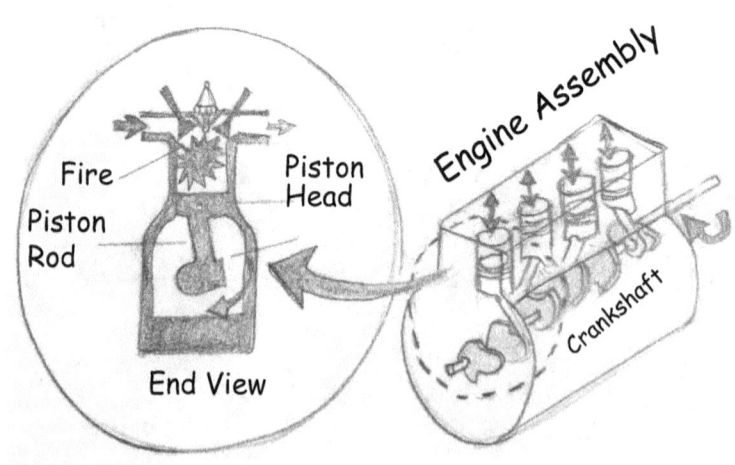

Fire

Piston Head

Piston Rod

Engine Assembly

Crankshaft

End View

For a copy of this video contact:

When you watch this video, keep in mind the seven steps to make cars.

Cars

. Get the materials for
the "Do" Demonstrations.
— Cookies Come From
— Clothes Come From
— Mold Chocolate
— Reverse Car Make
— Electricity Examples

. Printout "Try It" Worksheets.
— 1) Look at Labels (Clothes)
— 2) Car Key Words

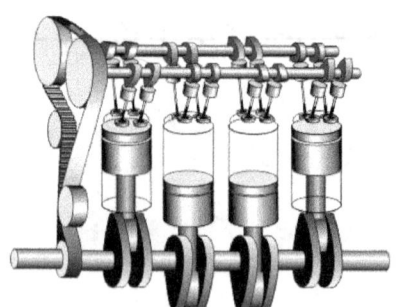

This lesson explains the
science of how to make
cars. The steps are:
pour, pound, cut apart,
push form, spot weld,
paint and finally put
together the parts.

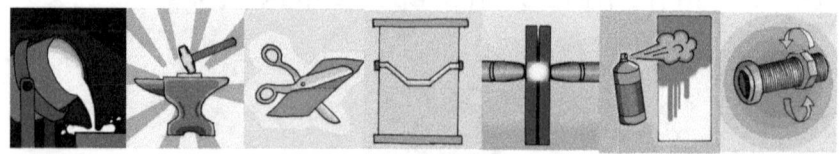

Before we see the science to make cars,
let's see how cookies and clothes are made.

— Cookies Come From

Purpose: turn ingredients into cookies

Materials: - see ingredient list
              - bowls, mixers, trays, oven

Steps:
1) Ask a parent to make cookies by video call.
2) Prepare the ingredients. Show to the class.
3) Mix together. Notice how the batter does not
              look like cookies yet.
4) Bake the cookies. Talk about how heat changes
              liquid batter into solid cookies.
5) Bring the cookies to school.
6) Discuss the steps it takes to make cookies.
7) Eat the cookies.
See the eBook — "Cookie Come-Froms."

STILL A FAVORITE TODAY

ORIGINAL NESTLÉ®
TOLL HOUSE®
CHOCOLATE CHIP COOKIES

2 1/4 cups all-purpose flour
1 teaspoon baking soda
1 teaspoon salt
1 cup (2 sticks) butter, softened
3/4 cup sugar
3/4 cup firmly packed brown sugar
1 teaspoon vanilla extract
2 eggs
One 12-oz. pkg. (2 cups)
NESTLÉ TOLL HOUSE
Semi-Sweet Chocolate Morsels
1 cup chopped nuts

Preheat oven to 375°F. In small bowl, combine flour, baking soda and salt; set aside. In large mixer bowl, combine butter, sugar, brown sugar and vanilla extract; beat until creamy. Beat in eggs. Gradually add flour mixture. Stir in NESTLÉ TOLL HOUSE semi-sweet chocolate morsels and nuts. Drop by rounded measuring tablespoonfuls onto ungreased cookie sheets. Bake at 375°F, 9-11 minutes. Makes about 5 dozen 2 1/4 inch cookies.

RICH & CREAMY

Fair Use

TOLL HOUSE

NESTLÉ   MORSELS

Cookies Come From — continue

a) Get the ingredients.

Sugar

Flour

Eggs

Butter

Chocolate

Ingredients are made of atoms.

b) Mix together.

c) Bake.

Teacher - Cars

Everyday Objects Have Science Inside

All around us are objects that
are made and work because of science.

Before we see about cars, let's look
in reverse, at how clothes are made.

We start with the completed clothing
and go backwards to the cotton plant.

See the eBook for how cotton becomes cloth.

— Clothes Come-From Demo

C)

E)

D)

B)

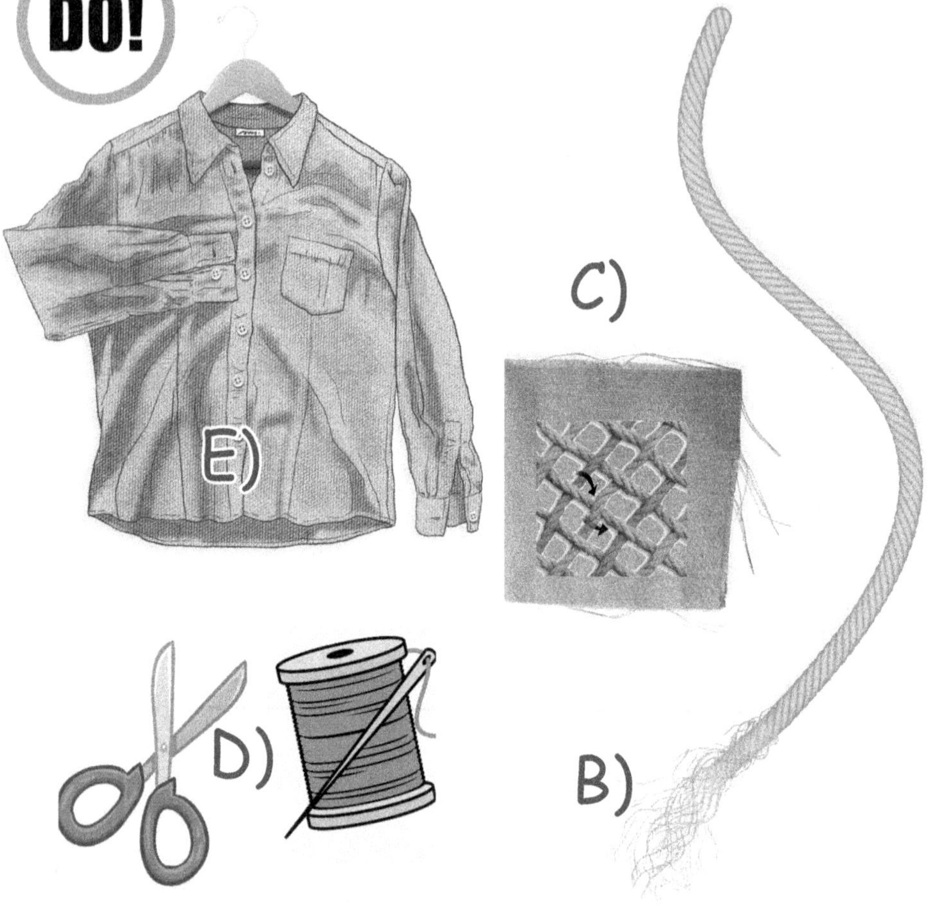

In Reverse
E) Clothes are made of cloth.
D) Cut and sew pieces of cloth.
C) Weave threads into cloth.
B) Spin fibers into threads.
A) Fibers grow on cotton Plants.

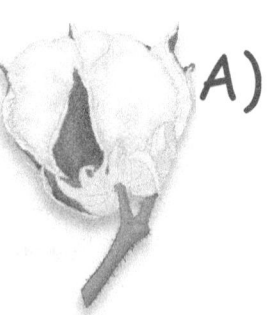

A)

Action steps that make things are called processes.

SCIENCE of CLOTHES

1) Twist Fibers

2) Color Threads

3) Weave Cloth

4) Weave Patterns

5) Color Cloth

6) Cut and Sew

7) More Clothes

Question (Q)
Where do clothes come from?

Answer (A)
Fibers twist into thread
that weave into cloth that
are cut and sewn into clothes.

Key Concepts
. Action steps (processes)
turn separate fibers into
colorful clothes.
. There are different types
of fibers like cotton, silk,
wool and chemical.
. Color can be added to
threads, cloth or clothes.

Look at Labels

Look at labels to see what fabrics are made from! What actions add colors? How are the fibers made into fabrics?

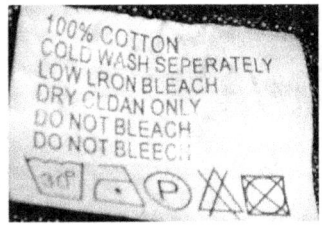

100% COTTON
COLD WASH SEPERATELY
LOW LRON BLEACH
DRY CLDAN ONLY
DO NOT BLEACH
DO NOT BLEEC.;

100% COTTON
MACHINE WASH
IN COLD WATER
TUMBLE DRY LOW
REMOVE PROMPTLY
NO BLEACH

50% COTTON
50% POLYESTER
MACHINE WASH WARM
TUMBLE DRY LOW
REMOVE PROMPTLY
DO NOT BLEACH
MADE IN U.S.A.

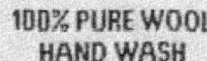

100% PURE WOOL
HAND WASH

100%
SILK
DRY CLEAN
ONLY

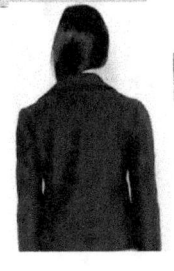

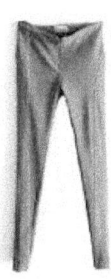

For a copy of this video contact:

Cars are made by actions called processes.
To better understand processes, let's look
at the actions that make clothes in 7 steps!

— Mold Chocolate

Purpose: mold hot liquid in different shapes

Materials: - chocolate & molds
              - heat source, containers

Steps:
1) Ask a parent to make mold
chocolate by video call.
2) Carefully, slowly melt the chocolate.
Teacher points out that things melt with heat.
3) Pour into liquid chocolate into molds.
4) Teacher discusses that it takes lots of heat
to melt metal. Liquid iron or aluminum is
poured into molds to make car engines.
5) Eat the chocolates.

Car Make Steps (Reverse Order)

Let's look at the steps that make cars (in reverse).

G) Put Together
F) Paint
E) Spot Weld
D) Push
C) Cut Apart
B) Pound
A) Pour

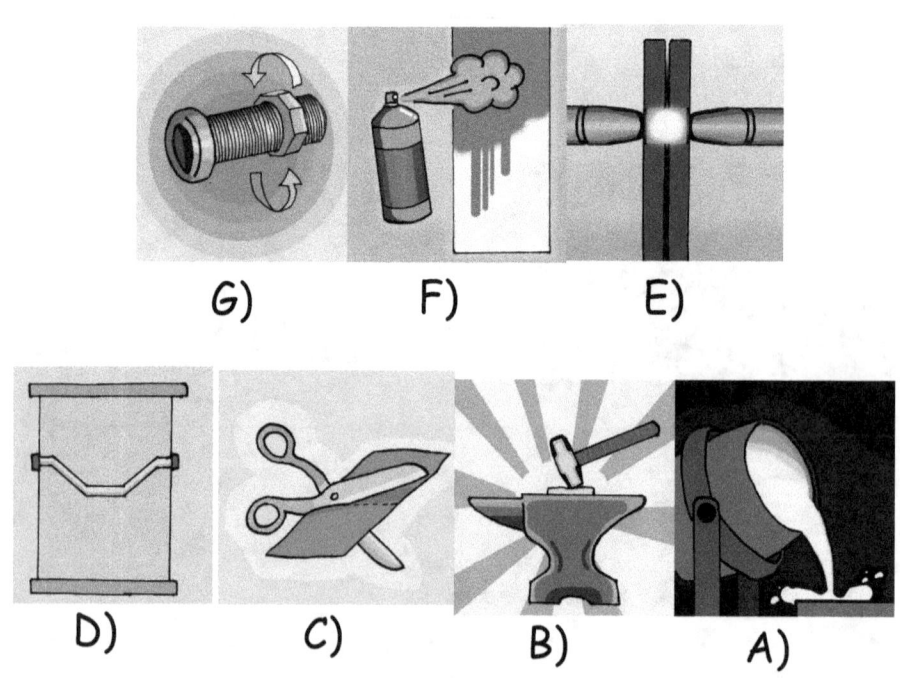

— Car Key Words

Question?

Match the key word with the right picture.

Cut Apart   Pour

Paint   Push

Pound   Put Together

              Spot Weld

A) _____

B) _____

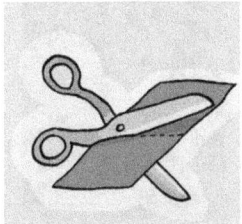

C) _____

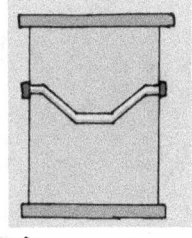

D) _____

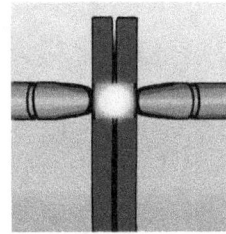

E) _____

F) _____

G) _____

TRY IT! — Car Key Words

Answers

Match the key word with the right picture.

A) __Pour__   B) __Push__   C) __Cut Apart__

D) __Pound__   E) Spot Weld   F) __Paint__   G) Put Together

F) RECAP

Science is in every step to make cars.
Pour the engine with heat. Pound crankshafts
with pressure until groups of atoms line up. Cut
apart gears with triangle tools. Push sheet
metal into molds with pressure to form frame
parts. Spot weld the frame assembly with many
small circles of electric heat.
Paint with charged spray. Finally, put
together the car with twisting torque
and pressing parts together.
Simply said, science with parts,
processes, machines and people make cars.

Cars — With 7 Make Steps

0) Heat and pressure are important to how cars are made.

1) Heat melts the metal that is <u>poured</u> into molds to make engines.

2) Pressure <u>pounds</u> hot metal to make crankshafts.

3) Wedge-shaped tools <u>cut apart</u> gears from room temp metal.

4) Pressure <u>pushes</u> sheets of flat, thin metal between tools to make car body frame parts.

5) Electricity heats spots of frame pieces to <u>spot weld</u> them together.

6) Negatively charged <u>paint</u> is attracted to the positively charged frame assembly.

7) All the pieces are <u>put together</u> to complete the car.

8) Cars go because gas burns in the engine. This causes heat and pressure. As the gas burns it puts pressure on pistons that turn the crankshaft. Gears transmit the turning to the tires.

G) Roll Up / Integrate Science
Cars — In 7 Steps

Heat melts the metal that turns into engine parts. With internal combustion, gas burns in the engine to push the piston and turn the crankshaft. Through gears and shafts the turning is transmitted to the tires to make the car move. DC electricity in car batteries turns into electromagnets that power motors to make electric cars move. AC electricity charges DC batteries to power our computers.

Engine

For a copy of this video contact:

When you watch this video, keep in mind the seven steps to make cars.

<u>Main Points</u>
1) Pour
2) Pound
3) Cut Apart
4) Push
5) Spot Weld
6) Paint
7) Put Together

Advanced
— Cars, The Commotion of Motion

Cars work because of the science of energy. Chemical energy in gas is burned in mini explosions inside engine cylinders that push pistons down.
The push is in sync to turn shafts and gears that then turn tires forwards and backwards on land and one day up into the sky.

Next time you are stuck in a traffic jam notice birds flying overhead. You can imagine what flying cars are going to be like.

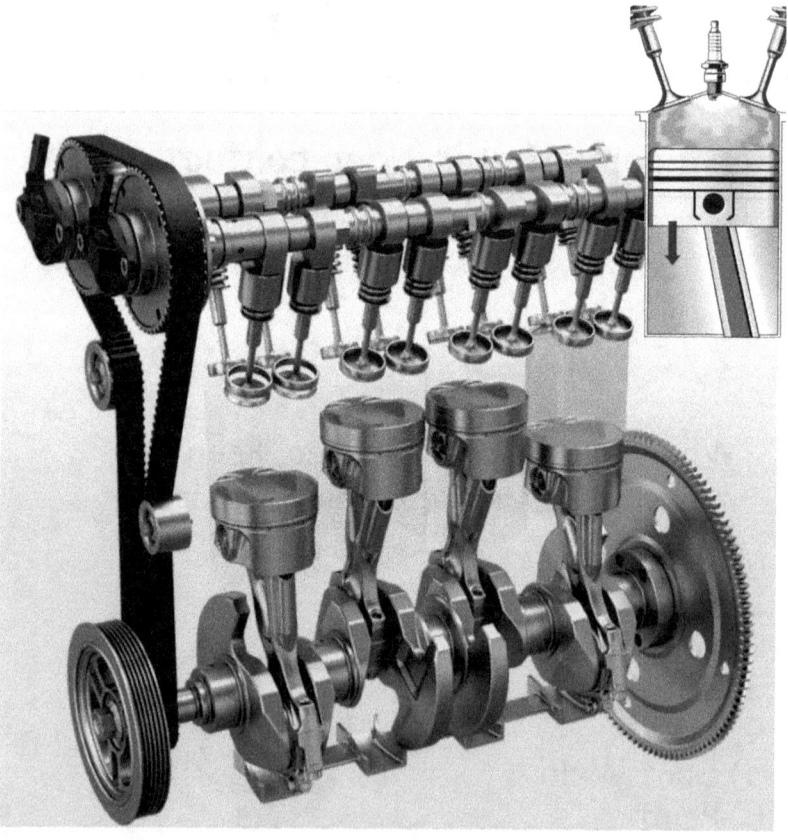

Science Story
— Molecules in Motion

Cars are atoms in action. Atoms and molecules are shaped with heat and pressure to make car parts that are assembled all together.

In the engine, molecules break apart, give off energy and then the atoms recombine to make new molecules. That is gas, (hydrogen and carbons atoms) joins with nitrogen and oxygen atoms from the air at the brief fire bursts inside the car engine. This pushes pistons down but also makes poisonous carbon dioxide and nitrous oxide.

The exhaust gases go into the catalytic converter where they change into CO_2, H_2O, nitrogen and oxygen.

As a side note, the first fuel spray carburetors use the same principle as perfume sprayers. Later, we will see how CHON atoms plus phosphorus are important to life.

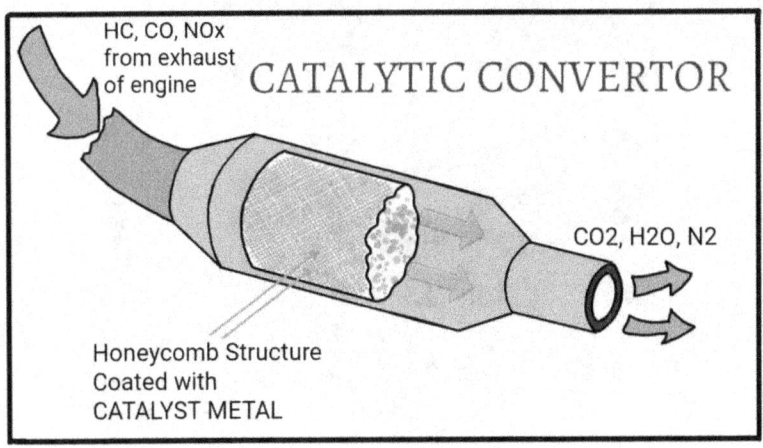

HC, CO, NOx from exhaust of engine

CATALYTIC CONVERTOR

CO2, H2O, N2

Honeycomb Structure
Coated with
CATALYST METAL

George Washington Carver
— Plant-based Car Fuel

"...the production of ethyl alcohol from home-grown products and waste will be our next successful venture"

—George Washington Carver, 1935
Birmingham News, from a letter to the editor
in response to "The Use of Alcohol as Motor Fuel"

Car Parts and Systems

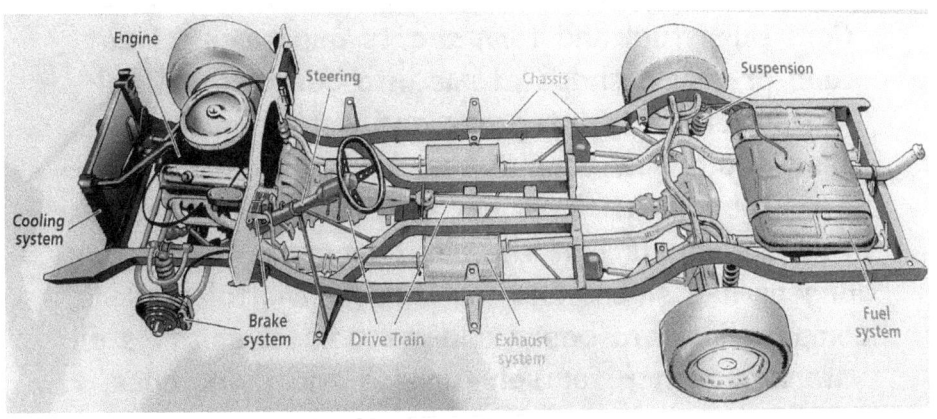

Fair Use.   LeMay Car Museum Tacoma Washington USA

Integrate

There is science with: how cars work, how they are made and what it takes to drive the cars. Cars burn fuel and turn shafts and tires. Heat and pressure shape atoms into car pieces. All the car parts integrate and work together. Think about how people drive cars. Humans sense the surroundings. Eyes see the road. Muscles turn steering wheels just the right amount. Ears hear dangers like sirens and flat tires. Digital sensors, computers, data bases and mechanical controls all work together for self-driving and flying cars.

Cars — 7 Make Steps

Factory — Clay to Cars

Power — Windows to Wheels

Teacher - Cars

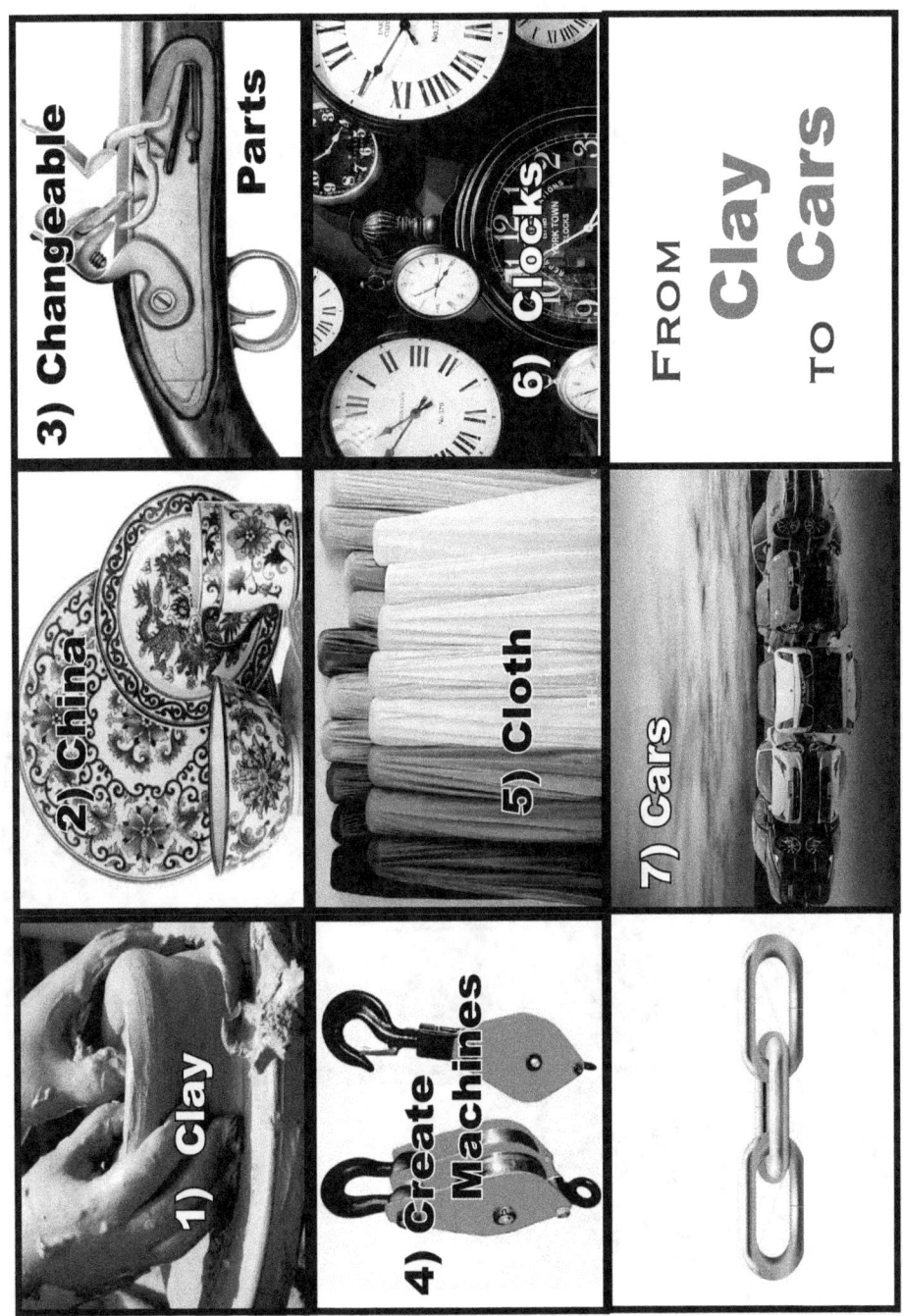

3) Changeable Parts

6) Clocks

FROM Clay TO Cars

2) China

5) Cloth

7) Cars

1) Clay

4) Create Machines

Factory — Clay to Cars

Purpose: Science connects factories from ancient clay soldiers to modern self-driving cars.

Main Points

1)   CLAY
Heat turns soft clay shapes
including soldiers into hard ceramics.

2)   CHINA
In the West, ceramic objects like plates and bowls
are called "china" for where they first come from.

Main Points — continue
— Factory — From Clay to Cars

3)   INTERCHANGEABLE PARTS
Mass-produced parts are practically identical. They assemble together to make our everyday objects.

4)   CREATE MACHINES
Throughout time people invent machines to replace muscle power with wind, water, steam, gas and electricity energies.

5)   CLOTH
Textiles are one of the first products mass-produced by machines and people in factories.

6)   CLOCKS
Time keepers with hands and gears are another early product mass-produced with interchangeable parts.

7)   CARS
Today, cars are made in factories by human engineered, computer-controlled machines in harmony with people workers.

May we look at our daily items with wonder and ask questions. What science is inside? How do they work? How are they made? Who made them?

Watch VIDEO
Factory — From Clay to Cars

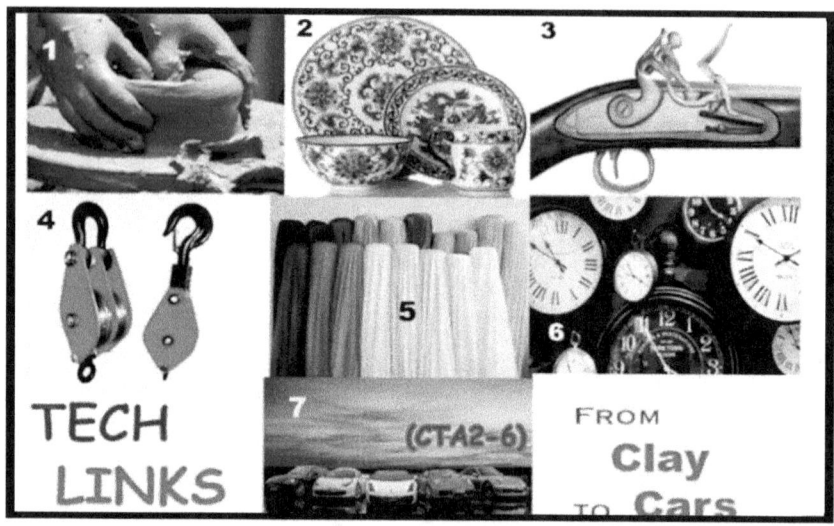

This is the true tale about factories. Tech connects from ancient armies of clay soldiers to modern cars. Our story starts over two thousand years ago.

For a copy of this video contact:

Advanced   Watch VIDEO
Science — Farms to Factories

For a copy of this video contact:

Main Points
1) Better Farms, More Food
2) More Iron, More Tools
3) Steam Power
4) Machines Make Tools
5) Gears (Machines Do Jobs)
6) Stronger Steels
7) Factories

Science Story
— Burned Basket, Baked Clay

The story is told of how ancient humans learn to weave thin strips of bark or stems to make baskets. At this time, grain is stored in woven baskets. Mice can eat through the baskets and eat the grain.

People coated clay on the outside of baskets and baked them hard in the sun. The clay encased baskets keep mice and bugs out of the grain.

One day by accident, a cooking hearth fire burns down a house. People notice that the plant material in the woven basket burns away but the clay actually hardens into a useful pot shape.

People learn how to bake clay with wood-fired kilns to make pottery. Not knowing that it is science, people experiment with clay and coloring materials. Each group creatively adds their own designs to their pottery. Clay pots store food, carry water and are even used to as cooking pots. Many people think that the need to make round pottery is why people create the first wheels.

 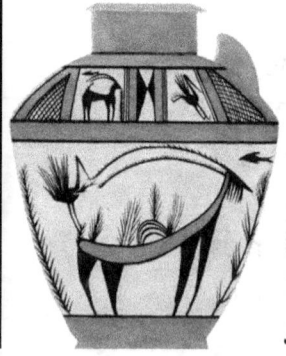 

Teacher - Cars - Factory

Integrate

Science integrates. Here is an example where geology, heat, pressure, fire and chemistry enable people to produce more iron and steel, and ignite global industry.

Over a billion years ago, microbes make oxygen and create Earth's atmosphere. The oxygen combines with iron and other elements to make iron ore that over time is buried. Next, prehistoric shelled micro-creatures settle on the floors of shallow seas. Over time, layers are buried in sand. With heat and pressure, they become limestone.
Millions of years ago, Earth's warm and wet climate leads to massive amounts of plants. Over time, they build up & decay. With time,heat and pressure they become buried coal.

Over geological time, the Earth's tectonic plates move on the Earth's surface. The land that is England today was once on the Equator, but it moved north.
During an Ice Age, thick, icy glaciers carve a valley called Iron Gorge. In the valley walls are exposed iron ore, limestone and coal.

Iron Ore

Limestone

Coal

Integrate

In 1709, Abraham Darby makes brass pots. He wants to use local iron, but impurities in the ore ruin the iron parts. He uses science to heat coal that removes the sulfur and phosphorus impurities. The heated coal is called "coke." It is a cost effective fuel. He makes a huge blast furnace with air-pumping bellows.

He uses the three local ingredients. Iron ore is crushed. Limestone is added because it purifies the ore. Coke has lots of carbon, so it burns well. Bellows add oxygen to the fire so it burns hot enough to melt the iron.

They pour the iron into shapes in molded sand.

The molten iron flows into shapes in the sand to solidifies. He makes and sells lots of iron pots. Next, he makes the outer cylinders for steam engines.

Later, people mass-produce steel for buildings, bridges, railroads and cars.

To summarize, nature made the ingredients. Glaciers carve out a valley, so people can see the iron ore, limestone and coal. With science, clever people energize an Industrial Revolution!

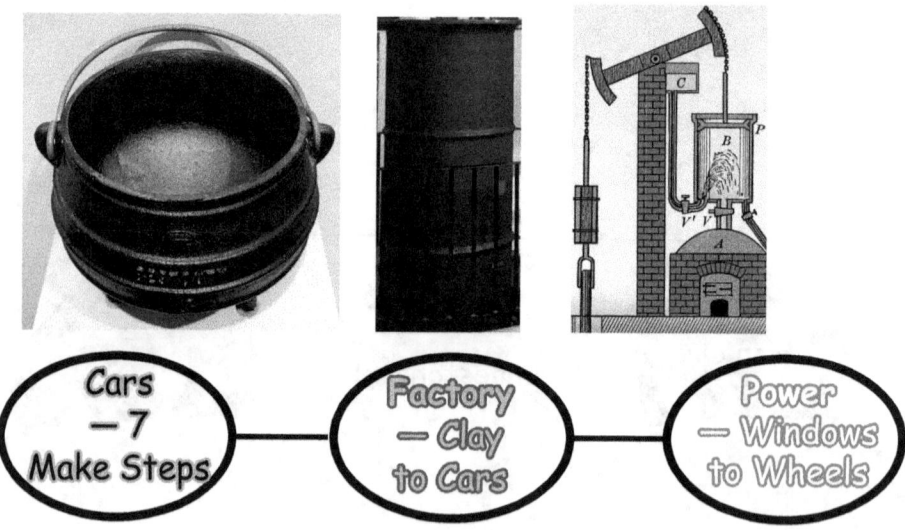

Cars – 7 Make Steps

Factory – Clay to Cars

Power – Windows to Wheels

Power — From Windows to Wheels

Power — Windows to Wheels

Purpose: Power is applied energy that we use to energize machines like cars.

Main Points

1) PIPE WINDOWS
Glass is heated sand, soda and lime. Some glass is blown in pipes into bulb shapes. In the past windows are made by spinning the hot glass bulbs into flat sheets for windows.

2) BREWER POTS
Beer makers heat ingredients in large pots called "vats."

3) PEAT COAL
Long ago, layers of dead plants become peat.
Over time, with heat and pressure it becomes coal.

4) STEAM PUMP
Brass beer boilers burn coal to heat water
to make steam that pumps water out of flooded mines.

5) IRON PRODUCTS
Coal is turned into cleaner coke to smelt iron
to make less expensive products including pots & cylinders.

6) STEAM POWER
The steam pump is improved to make
steam engines that power the industrial revolution in factories.

7) WHEELS PROPEL (Trains)
Boilers that hold higher steam pressure are
invented to power steam engines on iron tracks for trains.

Advanced   Watch VIDEO
Power — Windows to Wheels

(CTA3-9)

For a copy of this video contact:

Main Points
1) Pipe Widows
2) Brewer Pots
3) Peat Coal
4) Pumps
5) Iron Products
6) Steam Power
7) Wheels Propel

Tech connects with power from clear windows to moving wheels.
In between is beer, fires and then the story gets full of hot watery air called "steam."

Integrate

Power is energy in action with the following examples. We eat and turn food into ATP and then our physical energy. Plants turn sunlight into food.

Some prehistoric plants and animals over time turn into coal and oil. This is why we call them "fossil fuels." Steam engines burn coal to make steam to push pistons and turn wheels and shafts. Oil is heated to make different fuels for airplanes and cars.

Many power sources are used to make electricity. This includes: wind, water, fossil fuels and nuclear power that turn wire coils near magnets to make AC electricity.

AC electricity powers motors like our home appliances from air conditioners, blenders, ovens to refrigerators and televisions. Outlet AC electricity changes into DC electricity to power our digital devices.

Energy is able to change from one form into others.

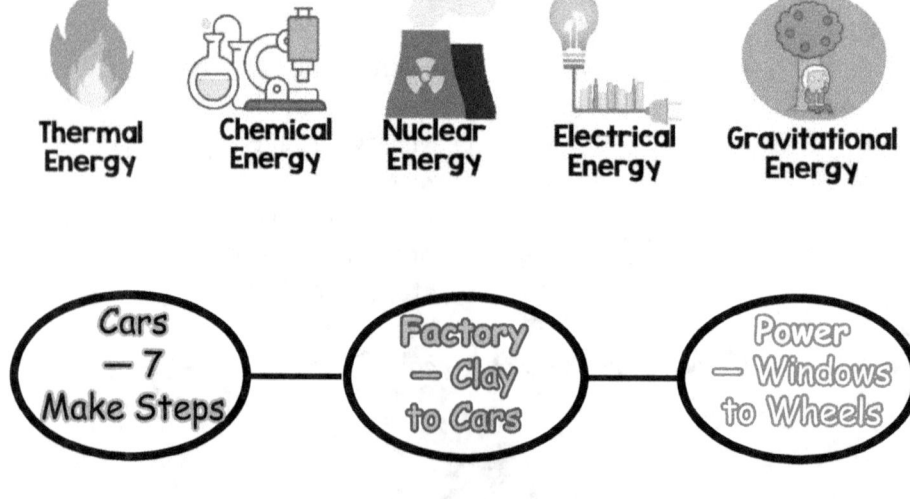

4

Advanced
— Electric Power

Scientists often share what they learn. This way others can build upon the discoveries and make inventions that improve lives. Electricity is one example. Volta makes batteries.

Oersted notices that electricity makes a compass needle move and figures out that electricity and magnetism are related.

Faraday experiments with this idea. He uses DC batteries to make the first electric motors. Later, he tries the reverse and discovers how moving a magnet near a coil of wire makes AC electricity. Today, electricity is still an important source of power for our everyday lives.

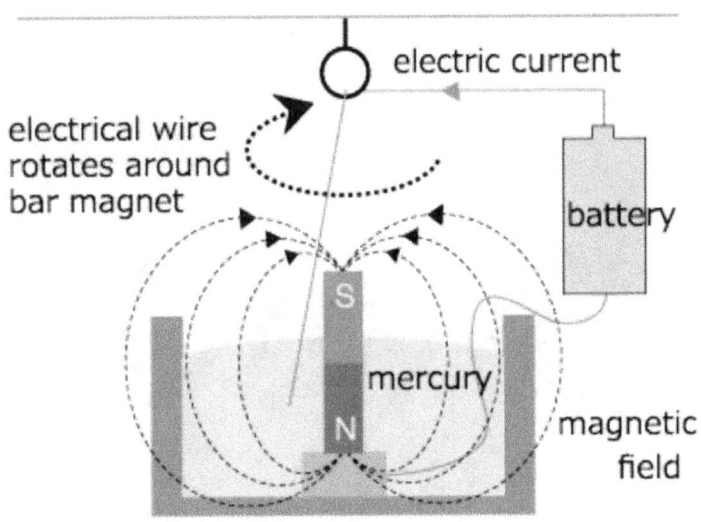

First motor that Faraday invented

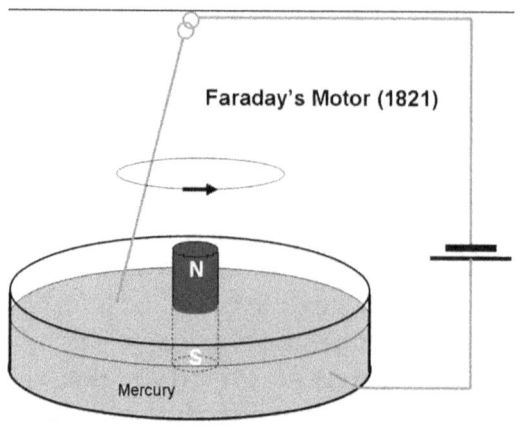

Faraday's Motor (1821)

Mercury

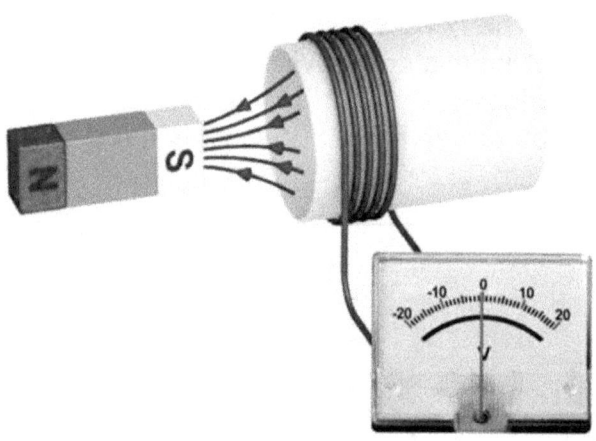

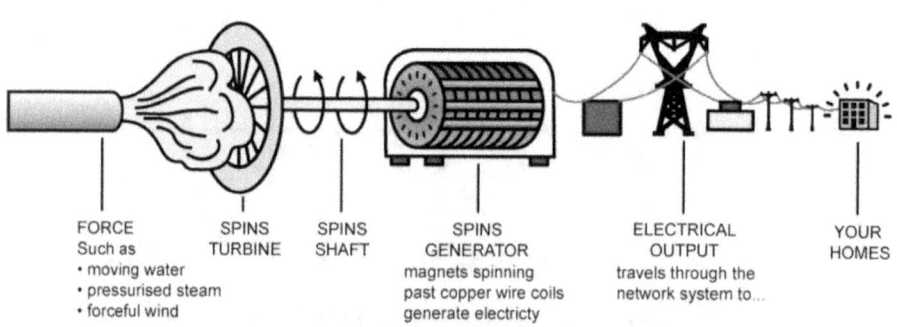

FORCE	SPINS	SPINS	SPINS	ELECTRICAL	YOUR
Such as	TURBINE	SHAFT	GENERATOR	OUTPUT	HOMES
• moving water			magnets spinning	travels through the	
• pressurised steam			past copper wire coils	network system to...	
• forceful wind			generate electricty		

Electricity

Electricity literally, powers
our world from plug-in appliances
to our connected digital devices.

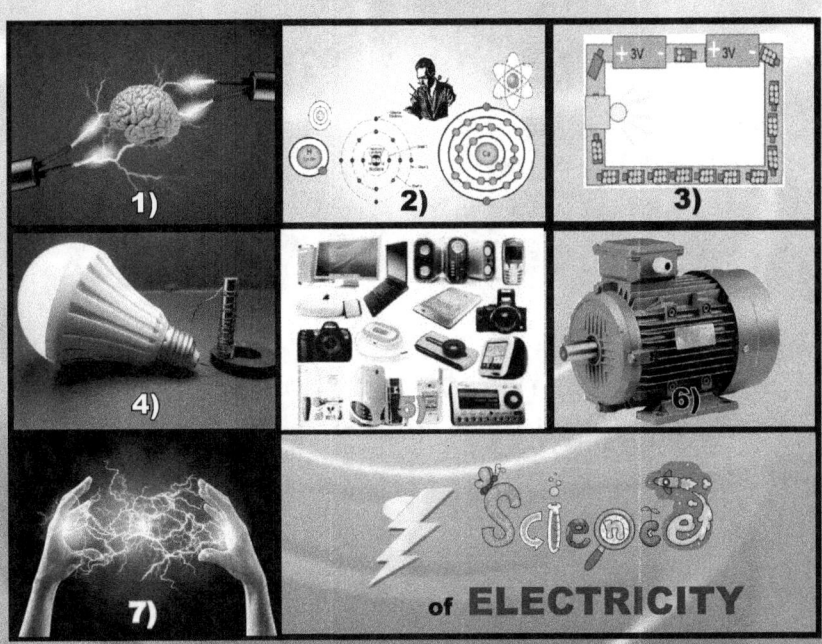

For a copy of this video contact:

Electricity zaps and shocks! It is the power in our digital devices. Let's learn more about electricity with 7 Ms.

Electricity — ONE PAGER

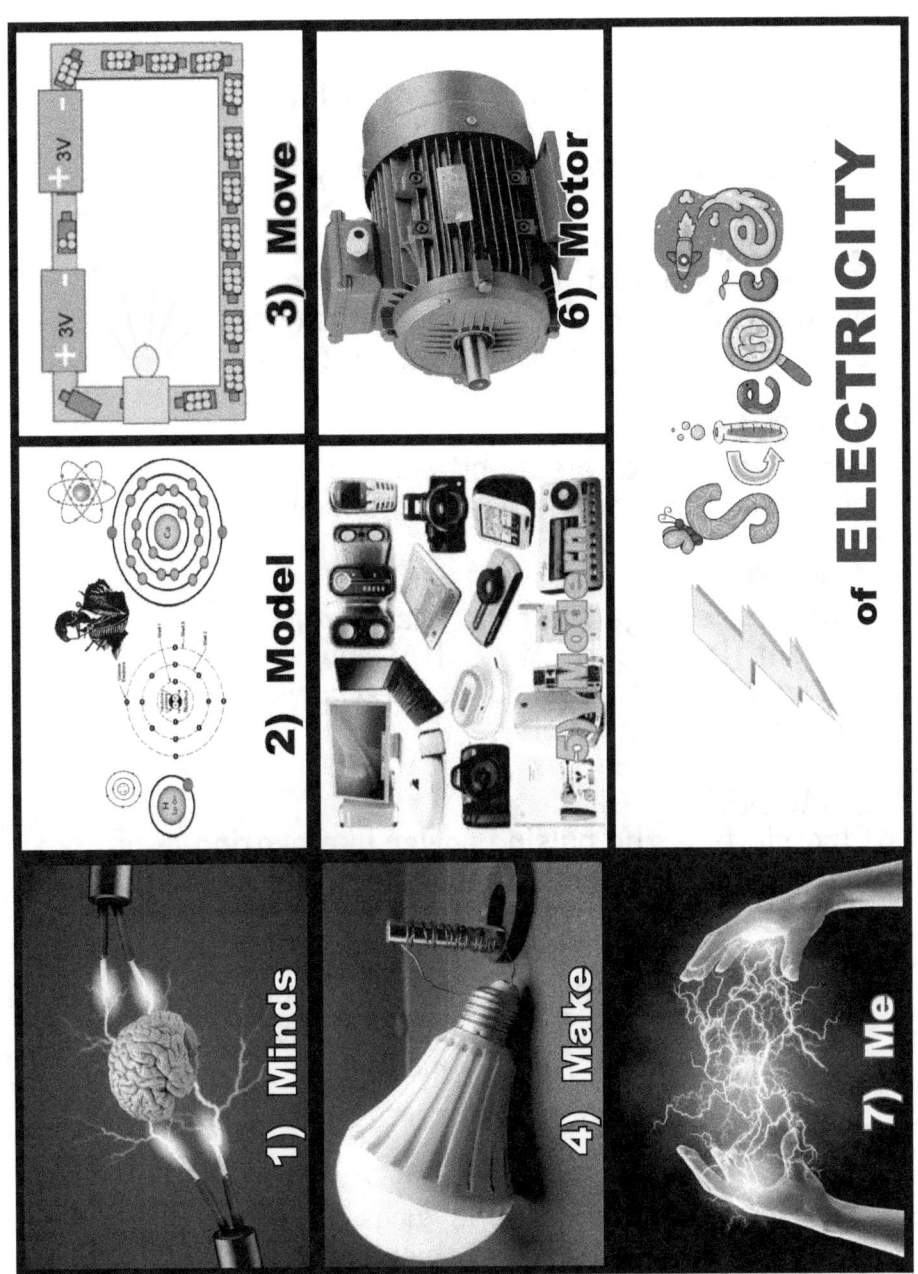

Electricity KEY CONCEPTS

Electricity is the oomph inside our electronics.
Let's look at electricity with 7 Ms.

1) MIND
People notice electricity in nature like
lightning and electric animals. People
think about how to make electricity.

2) MODEL
People experiment and create a model of
an atom. Electrons with negative charges
orbit above the positive center nucleus.

3) MOVE
People learn that electricity is
electrons (electric fields) on the move.

4) MAKE
Batteries change chemical energy into direct
current (DC) electricity. Also, moving a coil
of wire near magnets makes alternating
current (AC) electricity.

5) MODERN
Electricity is the pulsing power like beating
hearts that enables our modern digital devices.

6) MOTOR
Much of our modern world moves because
of electric motors from home appliances,
some cars to most factory machines.

7) ME
There is electricity inside each of us. Biochemical
electricity powers our hearts, lungs and senses.
Our minds use electricity to think , control
muscles and ponder exciting topics like electricity.

— Electricity Examples

Teacher to discuss the importance of
safety and electricity. Demonstrate battery-
powered devices like toys, DC motors etc.
Demonstrate AC-powered devices like a hair dryer
plugged into an outlet. Discuss AC motors.
Demonstrate how AC electricity can be changed
into DC. Show plugging in a digital device
like a smartphone and its charger. AC
electricity from a power plant flows from the
outlet and charges batteries in digital devices.
Electric cars have DC batteries charged by AC.
Class to discuss different examples of appliances
and digital devices powered by electricity.

Sky Electricity Science

We see electricity in nature. Sky e-Bits travel to Earth and turn into L-Bits of lightning and sound waves of thunder along the way.

Electricity Science

At home, our outlet AC electricity powers motors.

washing machine   refrigerator

hair dryer   vacuum cleaner   electric fan

Electricity Science

Electricity is inside our electronics. Input AC
e-Bits change into DC e-Bits to power our digital devices.

Electricity Examples

Search the Internet for examples
of electricity and how it is used.
Cut and paste or draw pictures below.

F) RECAP
<u>Electricity — With 7 Ms</u>

1) Human <u>minds</u> notice and discover electricity.

2) People <u>model</u> atoms with electrons that flow and cause electricity.

3-4) Our muscles and electric <u>motors</u>, <u>move</u> because of electricity.

5) People learn how to <u>make</u> DC electricity with batteries and AC electricity that flows from outlets.

6) Our <u>modern</u> digital devices use electricity. It is why they are called electronics.

7) Electricity is in <u>me</u> (each of us) also. Actually, it is inside everything that lives. Our hearts beat, senses sense and minds think with electricity!

Batteries have steady, direct current (DC)
electricity. Outlet electricity is alternating current
(AC) which means the current flow changes
direction over 50 times a second. AC outlet
electricity charges our smartphone batteries.
Our digital devices work with DC electrical e-Bits.
Inside smartphones, DC electricity changes
into AC electricity that makes radio waves.
Home satellite dishes also receive radio waves
that are changed into e-Bits for our TVs.
On our digital screens, we watch movies.

Advanced — Train Science —
Watch VIDEO

(CTA2-10)

For a copy of this video contact:

Let's look at train
science with 7 S's.
<u>Main Points</u>
1) Steam Engines
2) Steel Tracks
3) Slide Circles
4) Survey Path
5) Build Steps
6) Run Schedule
7) Impacts to Society

3) Cars
—Teacher Guide—

Heat and forces in action, pour, pound and push atoms into autos.

Science

EDUSTORE AFRICA

Indě Ed Project
Charitable Orgn

3) Cars
— Actions to Autos

3) Cars

Science pours, pounds and pushes metal and plastic into car parts. Cars have to be made before they can move us around!

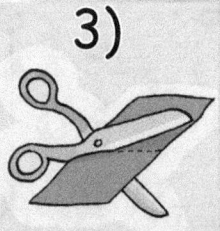

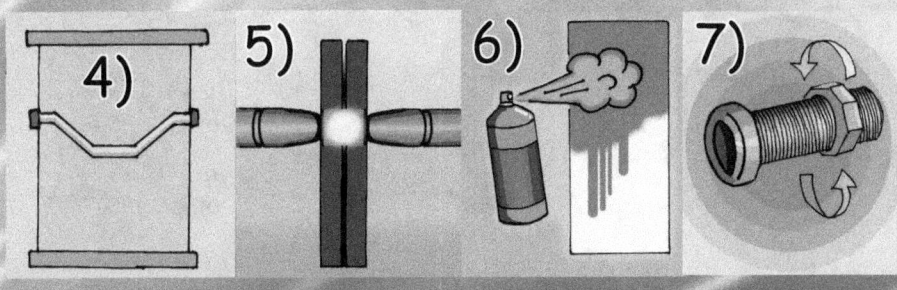

3) Cars!
– Actions to Auto

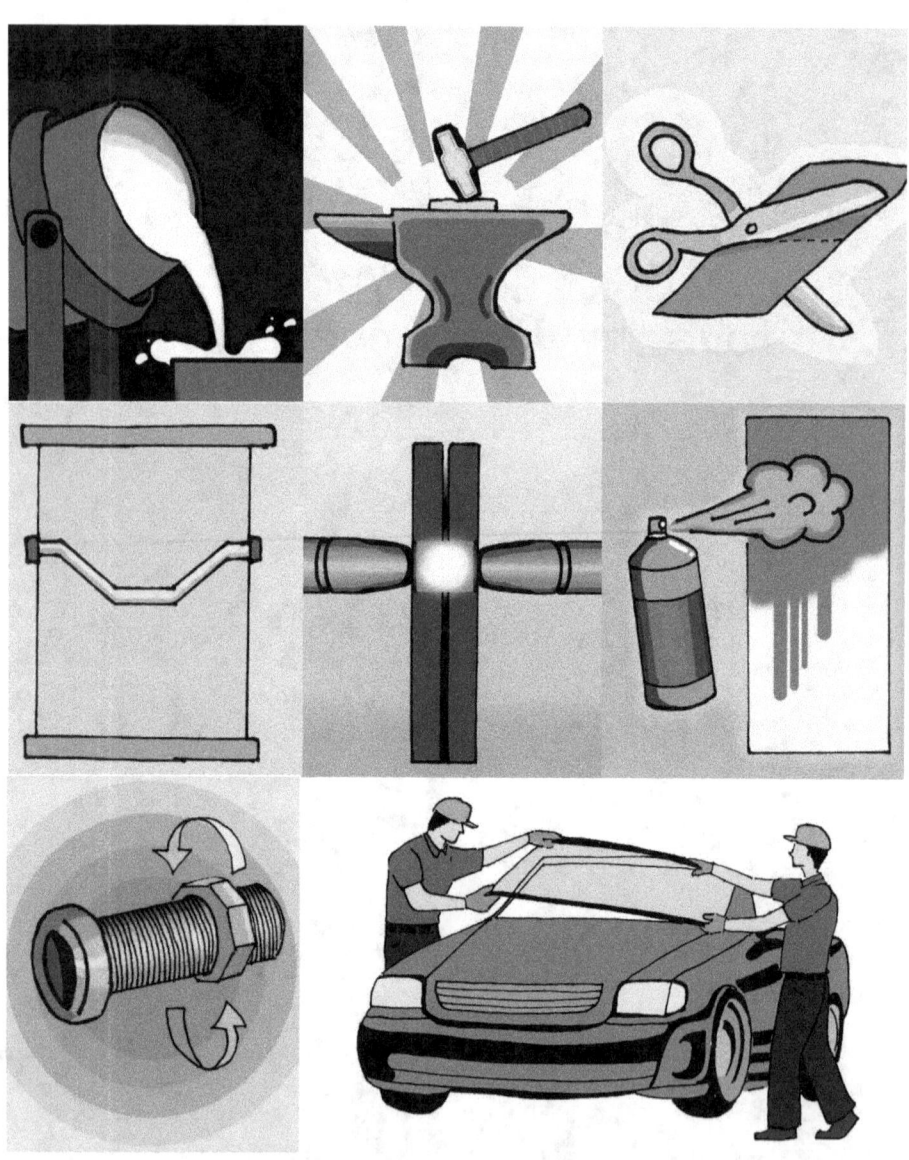

Seven MAKE Steps

Cars!
– Actions to Autos

This is the true story of
how science makes cars.
. From Atoms
. Then Piece Parts
. To the Complete Car

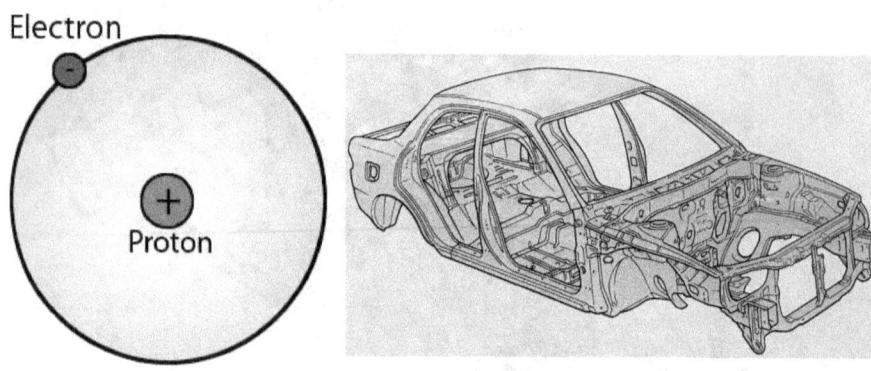

Cars!

Table of Contents

Cars!
— Seven MAKE Steps

Table of Contents

Cars – MAKE Summary

1) Pour
2) Pound
3) Cut Apart
4) Push
5) Spot Weld
6) Paint
7) Put Together

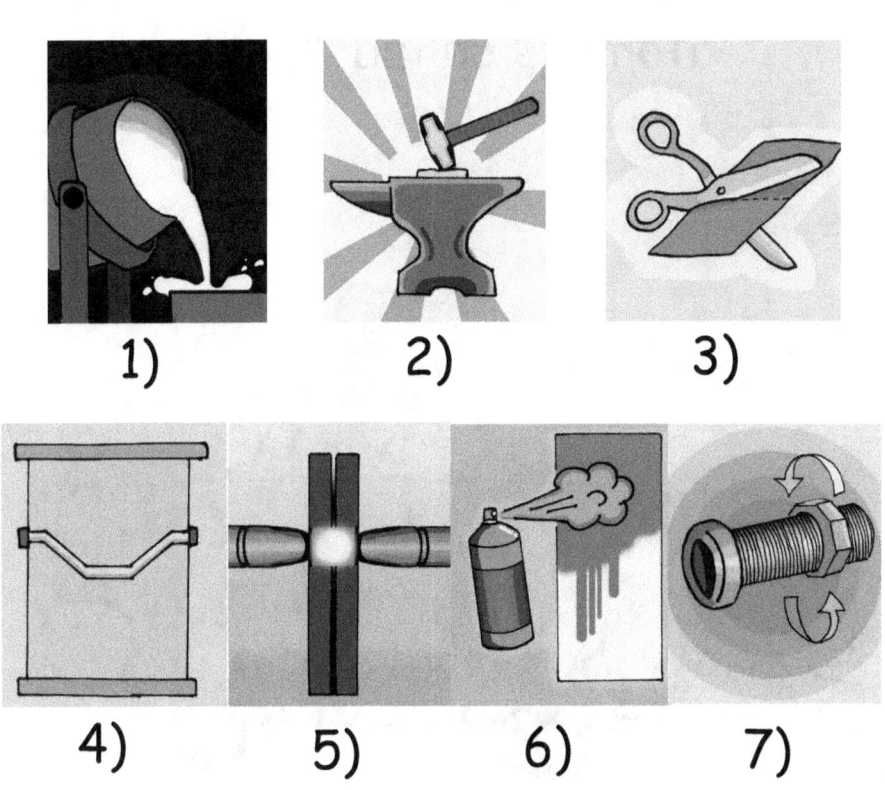

1) 2) 3)

4) 5) 6) 7)

Start Small

Before we make cars,
here is some simple
science. Parts are made
of very tiny atoms.
As a matter of fact,
anything made of
atoms is called "matter."

matter

Charges

Inside atoms are two charges,
positive (+) centers circled
by negative (-) electrons.
Electrons flow to make
electricity.

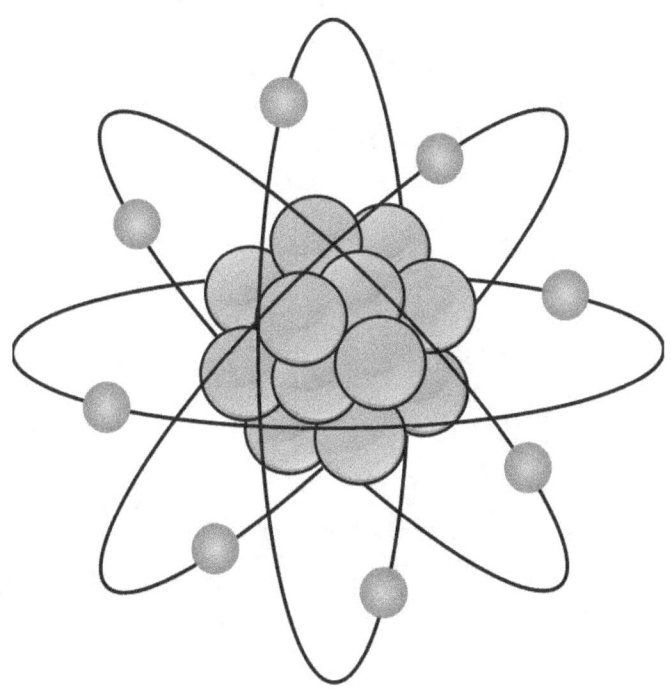

Next, we learn more about
atoms, molecules and mixes.

Atoms

Different atoms
are made of different
numbers of positive
and negative parts.
Atoms that are all
the same are called
elements. Iron is
an example.

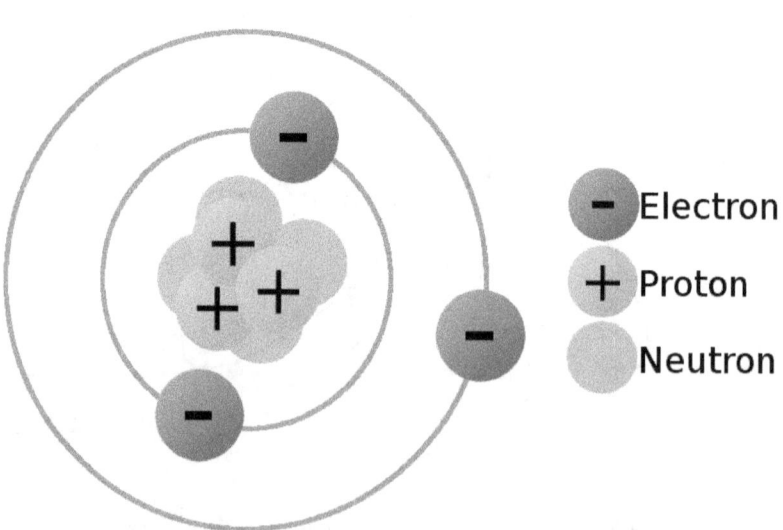

Electron

Proton

Neutron

This model is the lithium atom.

Here are some more atoms.

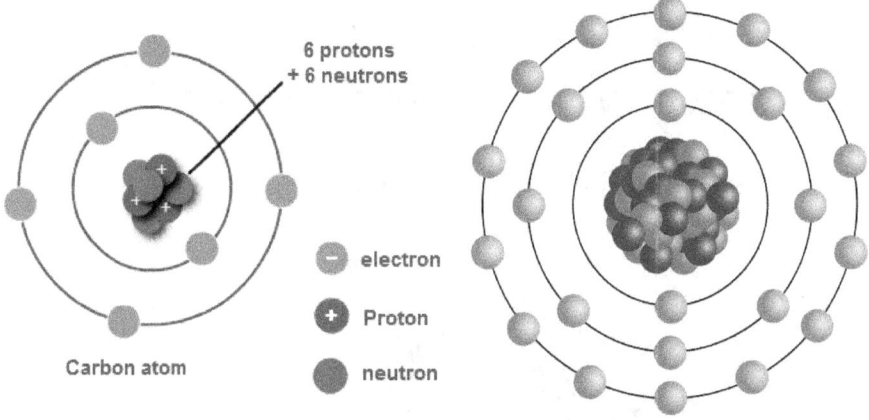

6 protons
+ 6 neutrons

electron

Proton

neutron

Carbon atom

Carbon atoms
have 6 + and
6 - parts.

Iron atoms
have 26 + and
26 - parts.

Atoms can join together.

Neutrons are not talked about in this book.

Molecules

Different atoms join together to make molecules. Water is an example. Two hydrogen atoms join with one oxygen atom to make one water molecule.

Water is also called H2O.

Mixes

Atoms also mix together like a fruit salad to make new materials. Steel is a mixture of separate iron and carbon atoms.

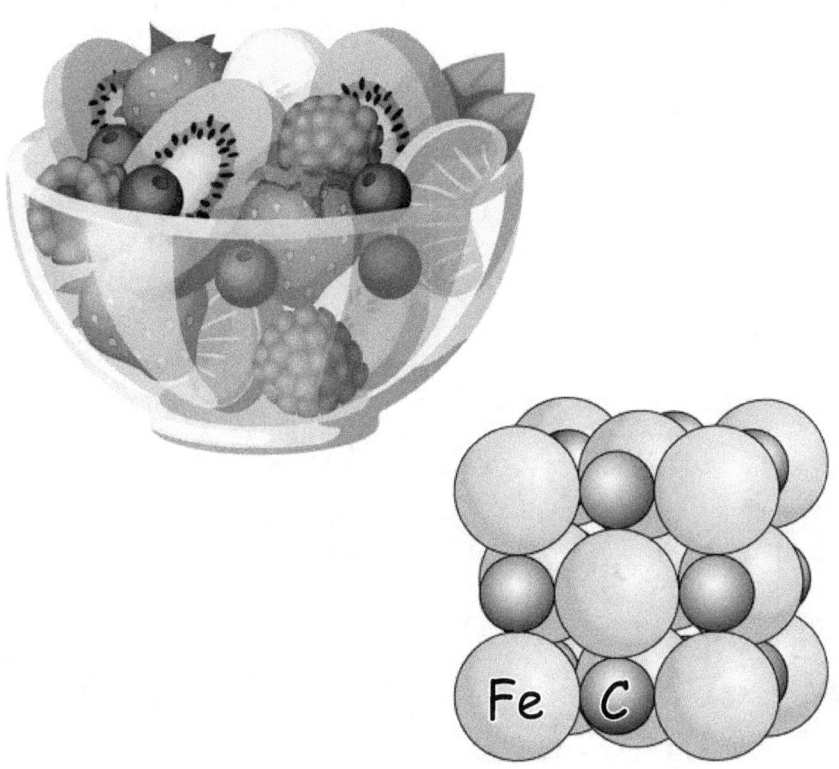

Fe C

Matter

Atoms, molecules and mixes are all matter. Matter comes three ways or states: solid, liquid and gas.
The state of matter depends on how much heat it has. For example: cold solid ice, warm liquid water and hot gas steam.

cold

hot

warm

Gas and Liquid

In a gas, atoms have space
between them. Air is a gas.
It is a mixture of mostly
nitrogen and oxygen.
In a liquid, atoms are closer
together. Liquids flow.
Tap water is a liquid.
So is molten metal.

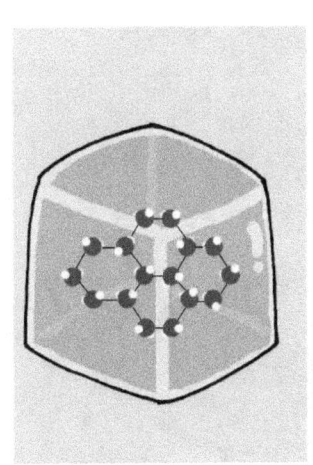  

"Molten" means to add heat
until something like metal melts.

Solid

When liquid atoms
cool, they line up
into a solid.
Ice cubes and metal
coins are examples.

Actual atoms are too small to see.

Cars

Heat and Pressure

Matter at different temperatures and pressures can be shaped into car parts.

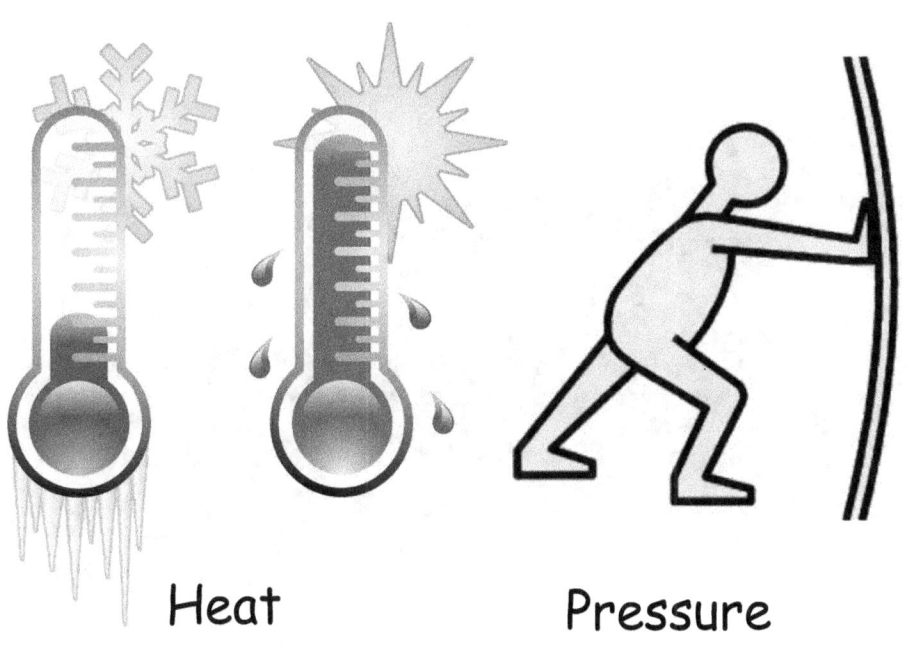

Heat   Pressure

Make

Now, we are ready
to drive on and see
how to make cars!

Cars are made of matter.
Cars are made by people and
machines including robots.

Processes

Let's learn about science and the seven steps or processes that make cars.

1) Pour
2) Pound
3) Cut Apart
4) Push Form
5) Spot Weld
6) Paint
7) Put Together

1) Pour — Cast

Pour or cast, very hot
liquid metal into a mold.

Engine

To make the engine, pour hot metal into a mold. The metal cools into a solid rough shape of the engine.

mold

Heat

Heat changes matter.
Water can be ice, liquid or
steam. Same with metals
but with higher temperatures.
Liquid metal cools into the solid
engine shape. Remove the mold
and it looks like this.

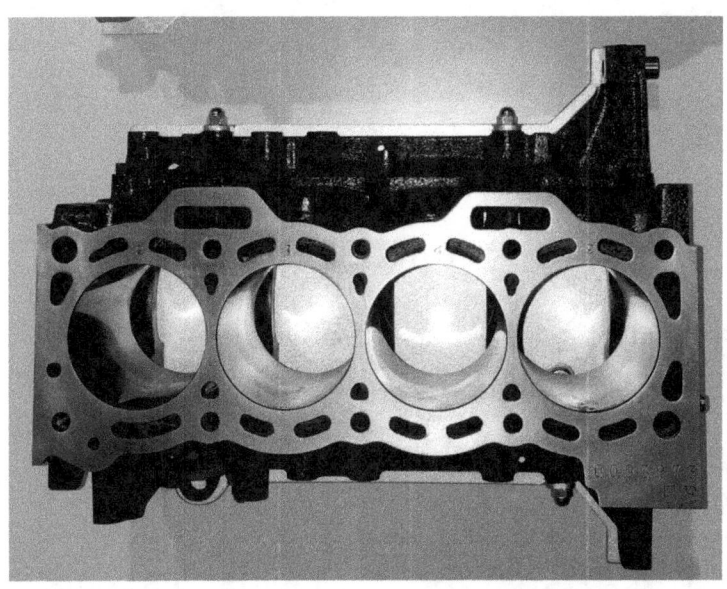

The big holes are for pistons.
Other holes, are for the cooling system.

More Cast Parts

Other car parts are cast too.

wheels

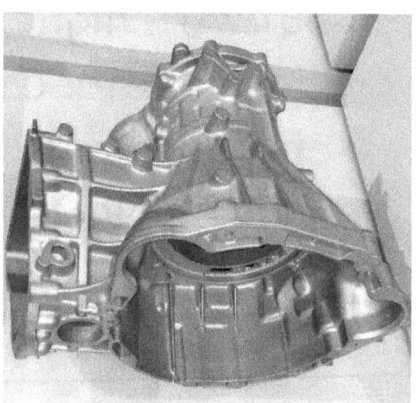

gear case

Like this steering wheel, many plastic parts are made by injection molding which is similar to casting. Hot liquid plastic is pushed into molds. It cools into the shape of the part.

Polish

Next, piston holes in the engine casting are polished to just the right size.

Casting makes parts of many shapes and sizes.
Castings are not strong enough to twist.
A new process is needed for parts that turn.

2) Pound — Forge

Pound or hammer to forge hot metal into the shape.

Crankshaft

The top tool pounds hot
metal into the under tool.
This shapes the metal
into a crankshaft.

top tool

crankshaft
(in work)

under tool

Tools forge the crankshaft.

Stronger Grains

Grains are groups of atoms.
They form when liquid metal
cools into a solid. In a casting,
atoms cool fast. The grains
are small and unaligned.
Forging forces groups of
metal atoms called "grains"
to flow into the shape of
the part. The aligned atoms
make the forged part stronger.

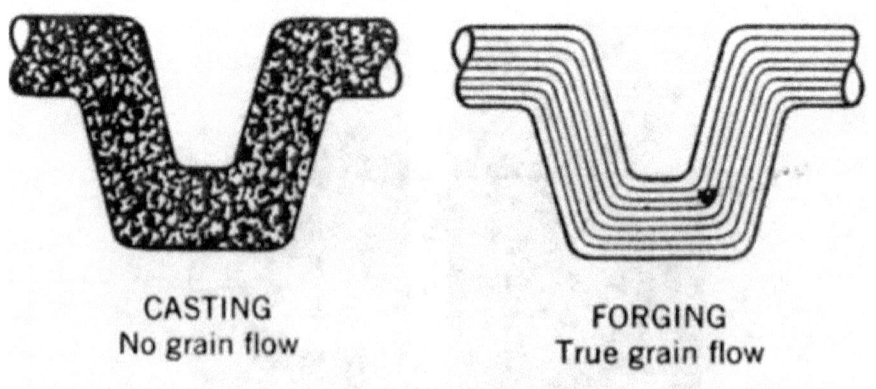

CASTING
No grain flow

FORGING
True grain flow

More Forged Parts

Piston rods are forged.

forging tools

Piston heads
are forged too.

Engine Assembly

The engine parts are put together or sub-assembled. This includes: the cast engine block; forged piston heads, rods, crank shaft etc.

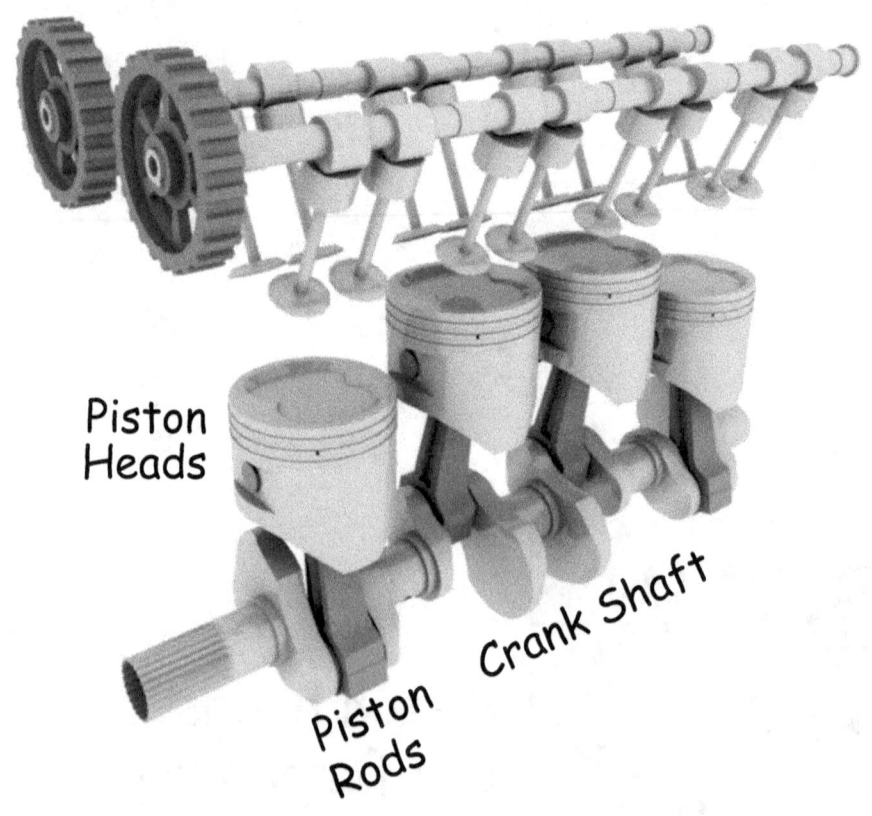

Piston Heads

Piston Rods

Crank Shaft

engine block not shown

Engine Action

Before we make more parts,
let us see how an engine works.
Gas burns in the engine. The
small, brief fires push the
pistons down. The pistons
turn the crankshaft.

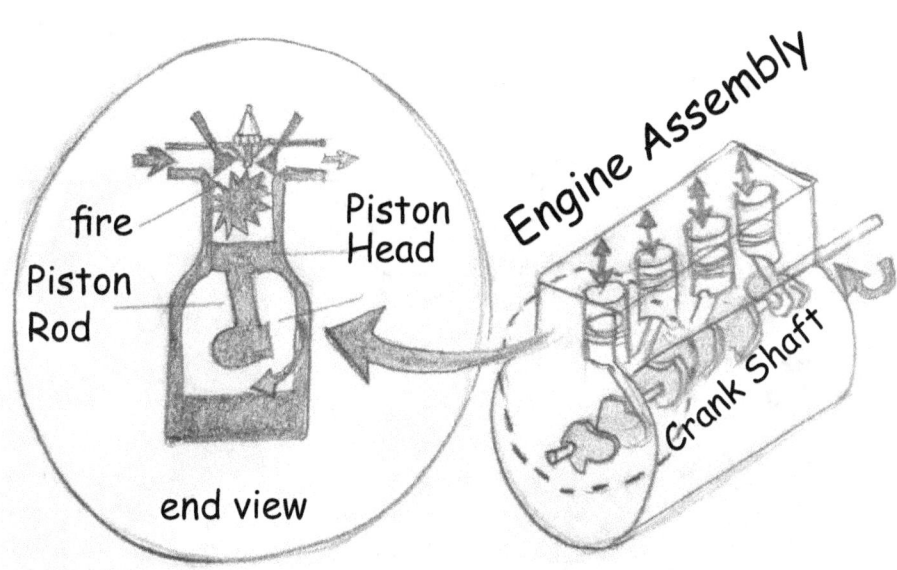

fire

Piston
Rod

Piston
Head

Engine Assembly

Crank Shaft

end view

More parts are needed to get
the turning power to the tires.

3) Cut Apart — Chips

Gears change or transmit turning. Gears are circles with teeth. Tools cut apart chips to make gear teeth. This is similar to scissors cutting paper or a knife peeling an apple.

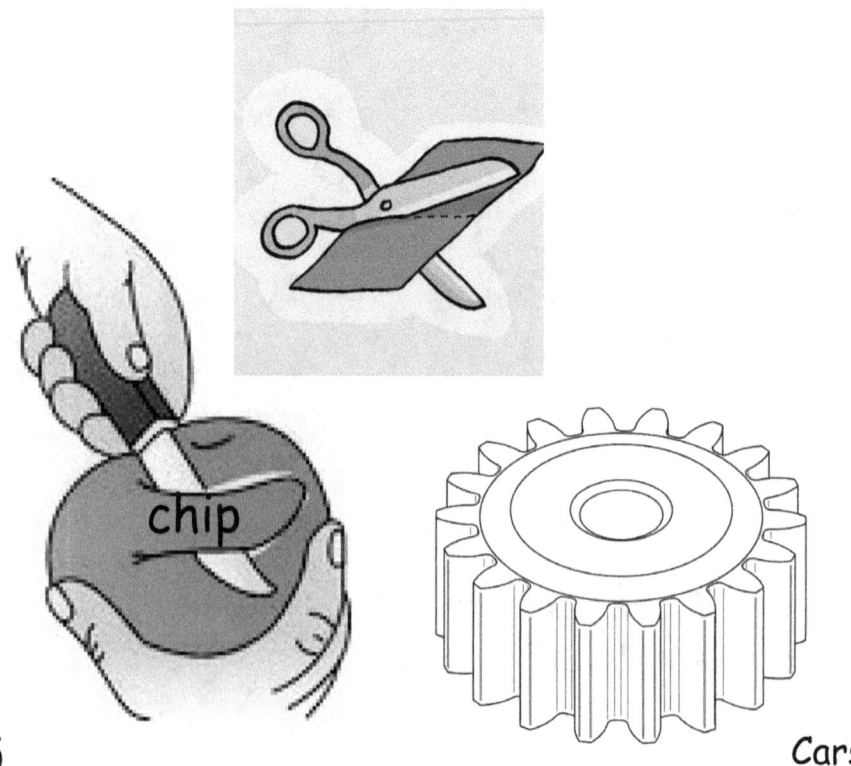

chip

Wedge Splits

The secret to making parts this way is triangle, wedge-shaped tools. Can openers, pizza cutters and axes are examples of wedges.

Cut Gears

Gears are cut apart from pieces
of room temperature metal
using wedge-shape cutting tools.

tool

gear

More Cut Gears

These gears change the power and speed output from the engine.

Synchro Assemblies

Change Direction

These gears change the
turning direction 90
degrees. That is, they
change the power from
the engine into power
for the tires.

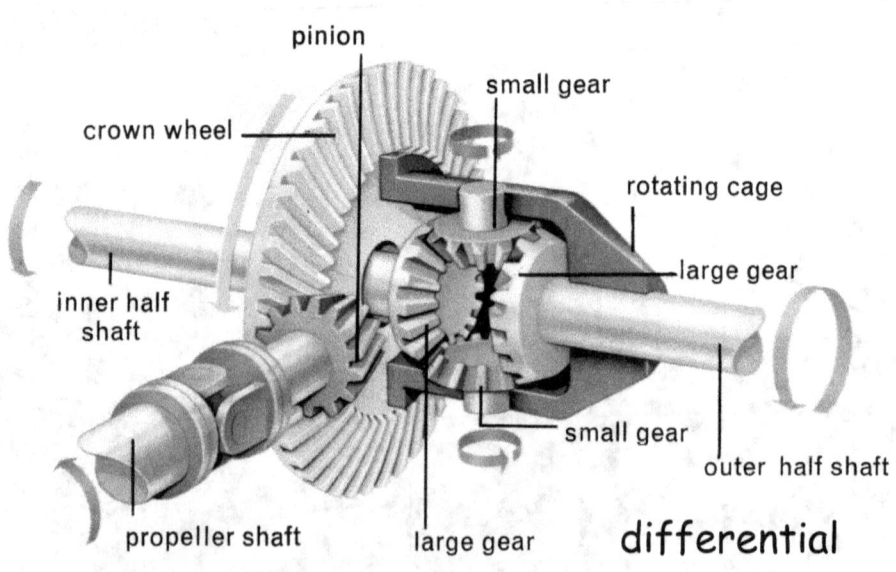

pinion

small gear

crown wheel

rotating cage

large gear

inner half
shaft

small gear

outer half shaft

propeller shaft

large gear

differential

But first we have to make the tires.

Press Tires

Layers of rubber, steel wire
and cloth are pressed together
in a mold to make hollow tires.

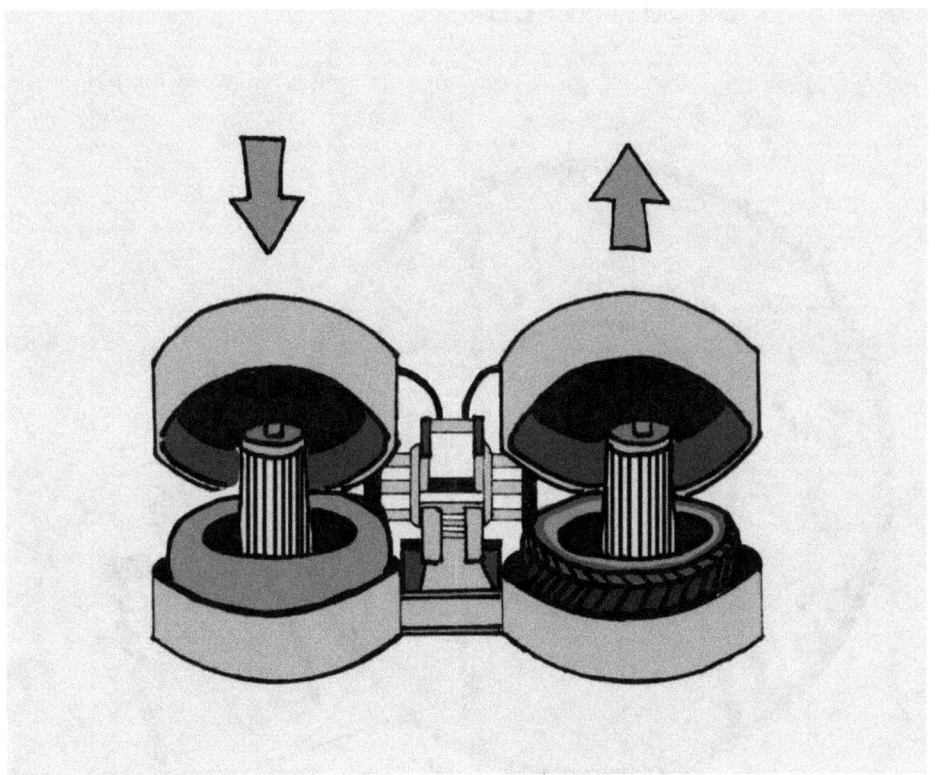

Air Pressure

Cars ride on tires full of air pressure. It makes the bumps in the road feel smoother for riders. Now, the car has parts that move. We need to make parts that stop the car.

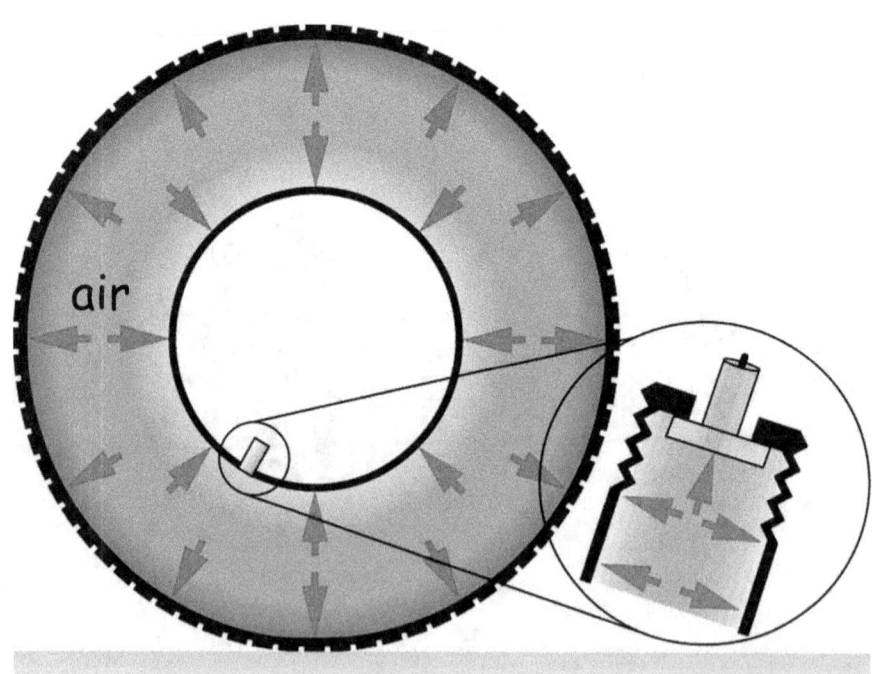

air

Brakes

Brakes stop the car. The brake system includes: pedals; pipes; pistons; brake fluid; pads and discs.

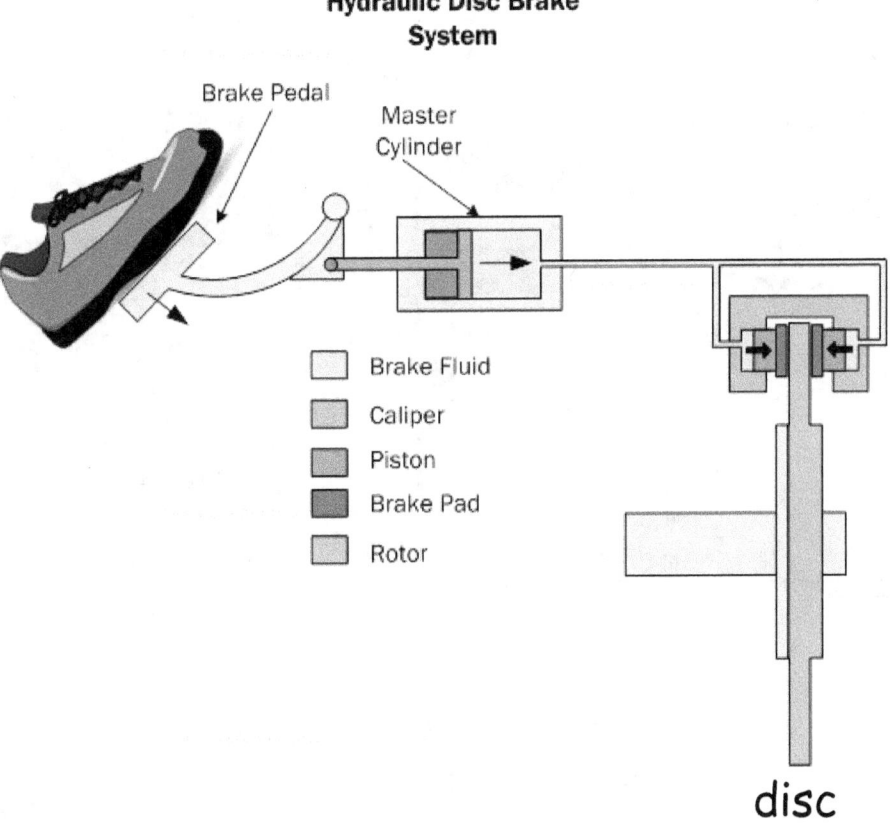

Hydraulic Disc Brake System

Brake Pedal

Master Cylinder

Brake Fluid
Caliper
Piston
Brake Pad
Rotor

disc

Press on the brake pedal
to put pressure on a liquid
in pipes and tubes.
The pressure moves from
the pedal to the brakes
to stop the car.

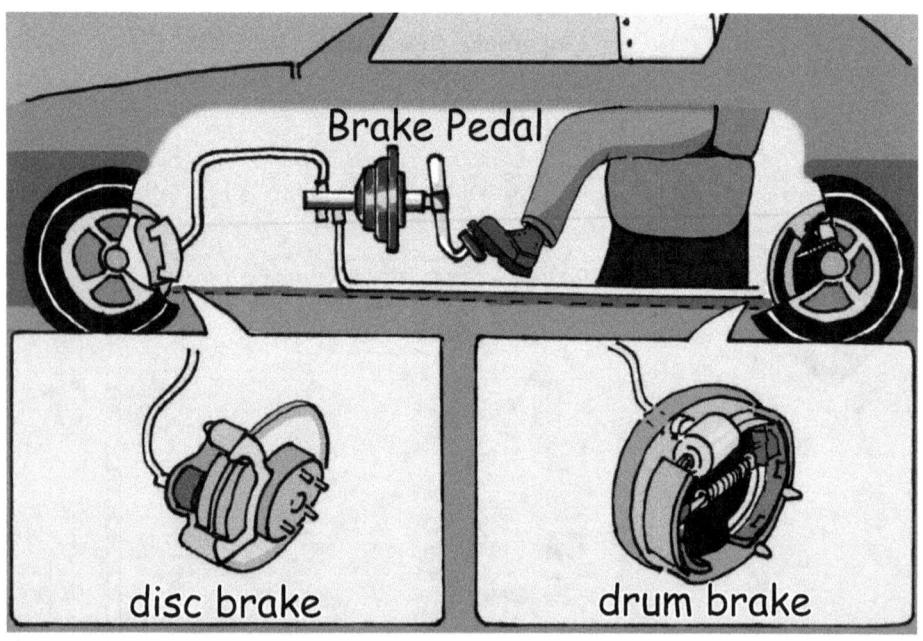

Brake Pedal

disc brake

drum brake

Next, we need a frame
to hold all the car parts.

4) Push Form—Frame Parts

Form is to push sheet
metal pieces into tools
shaped like car parts.

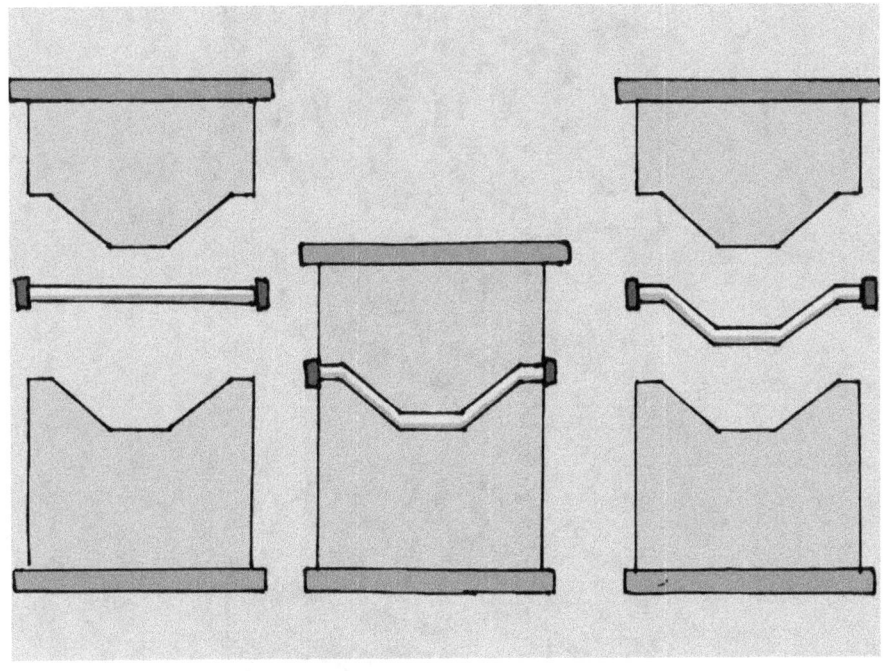

Sheet Metal Parts

The top tool, quickly presses
the warm sheet metal into the
under tool. This forms the frame
and other car parts like doors.

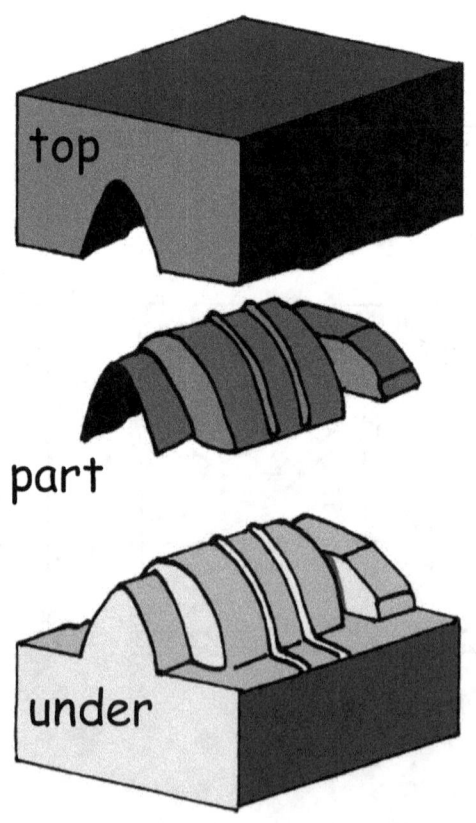

top

part

under

Pressure

The machine that makes big frame parts uses the weight of a **1,000** elephants.

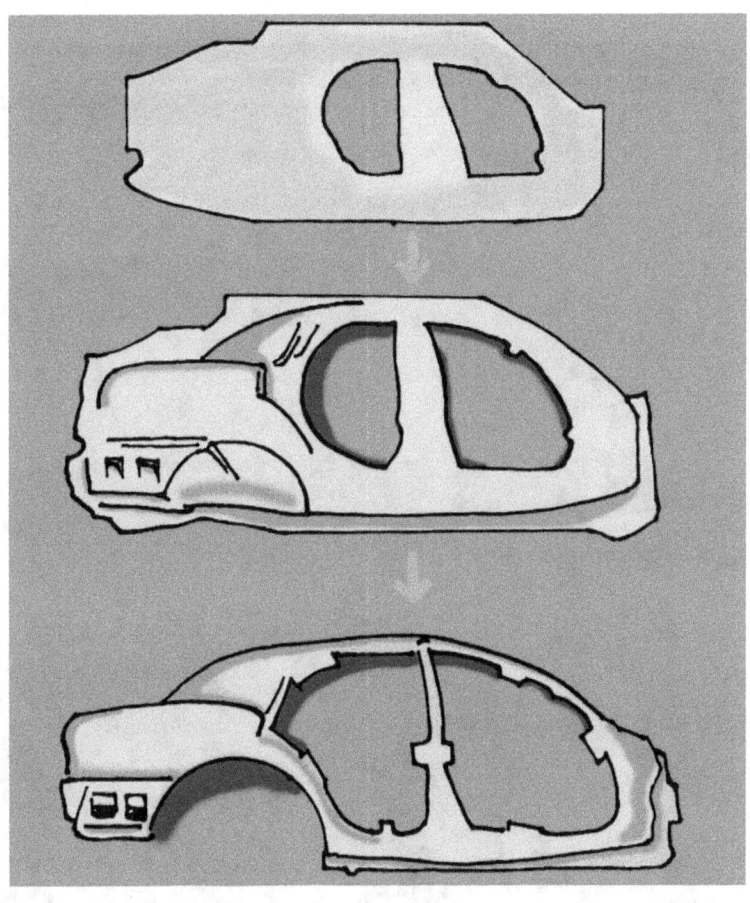

Pressure pushes the sheet metal into molds to make the frame parts.

More Push—Formed Parts

About half of the car's weight is made of push-formed parts.

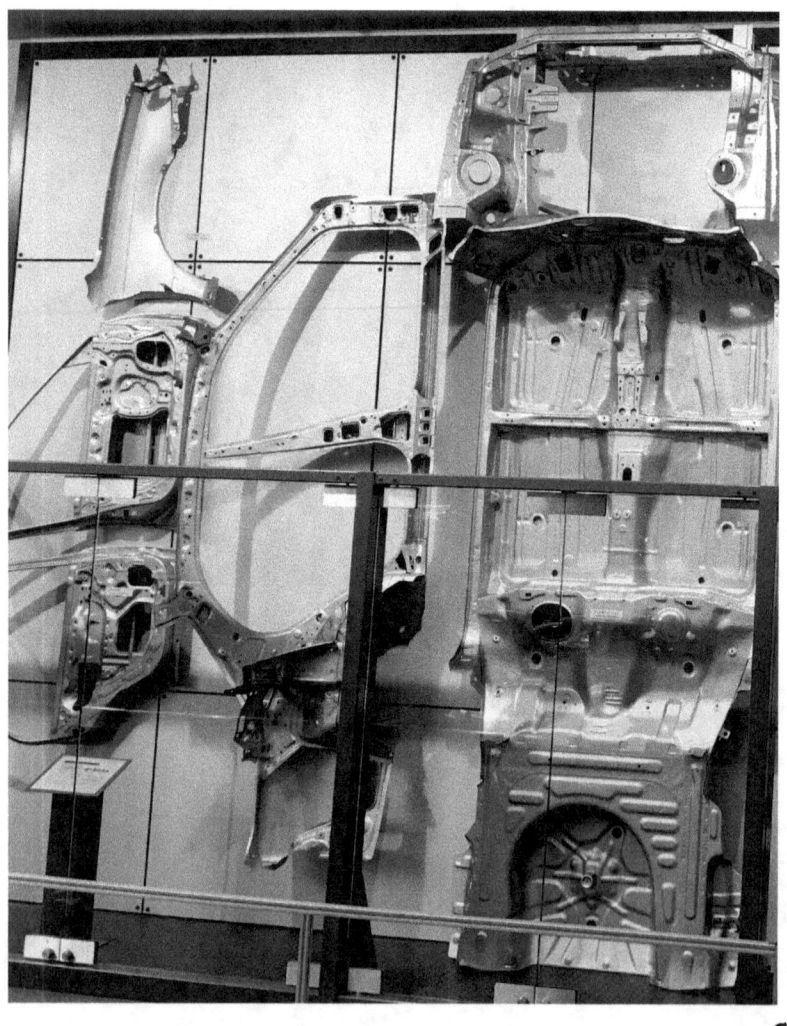

Next, we need to
join the frame pieces.

5) Spot Weld

Spot weld uses electricity
to melt points in metal parts.
The liquid hot spots, cool
and join the frame parts.

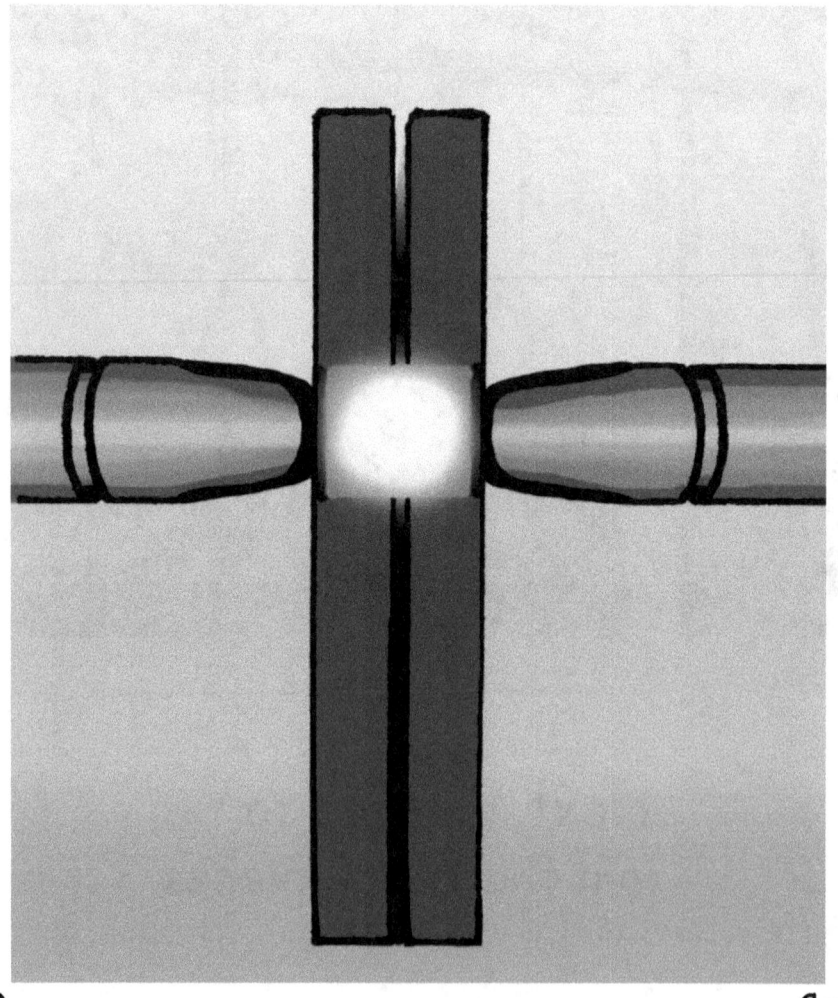

Frame Assembly

Robots spot weld sheet metal parts together to make the frame assembly.

Robots

That is, robots with finger-like tools clamp the sheet metal parts. Electricity flows and melts spots to join pieces together.
Robots make thousands of spot welds to join formed parts into one frame assembly.

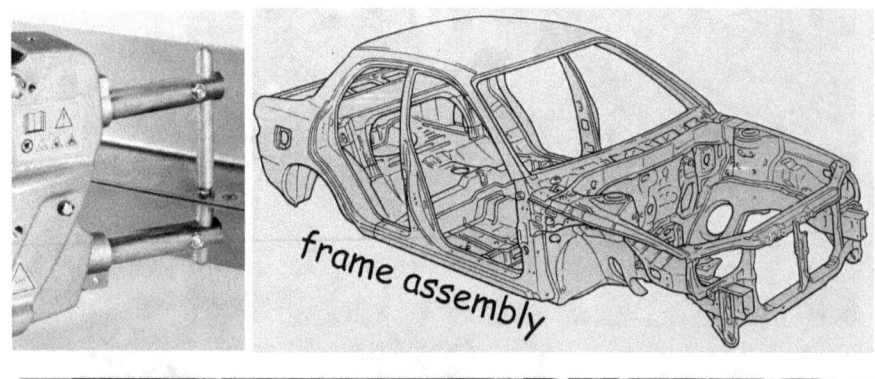

frame assembly

Electricity flows through robot weld heads.

Electrical Heat

Flowing electricity also makes heat. The electric heater, toaster and this light bulb are examples.

More Spot Welds

Other sheet metal parts like door assemblies are spot welded too.

Close-up of spot welds

Next, the frame needs color.

6) Paint

Paint the car in
three steps: 1) Under,
2) Middle and 3) Top Coats.

Under Layer

Paint is applied onto
the car in many layers.
First, the car is dipped
into undercoating. This
helps the car not corrode.

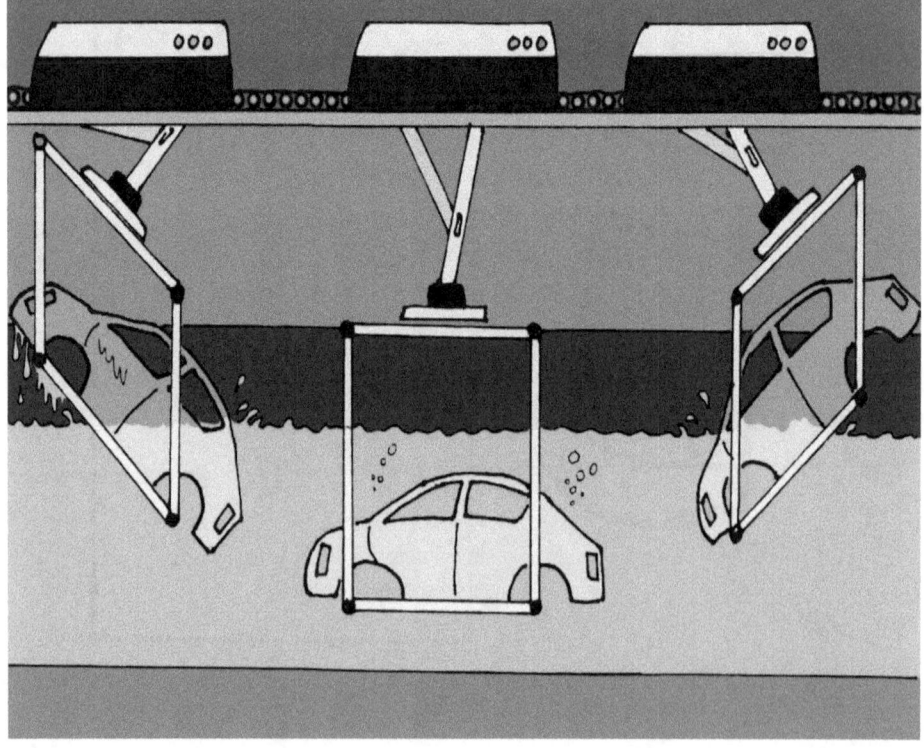

Middle Layer

Next, the middle coat
is sprayed onto the car.
This step fills in any
places that are not smooth.

Top Coat

Last, several layers of the top coat are painted onto the car. It is the top coat that gives a car its color.

Remember, inside atoms are two charges?

Charges

The paint is negatively
charged. The car
frame is positively
charged. That way,
tiny drops of paint
stick to the frame
at a smooth thickness.
The wet paint dries.
Excess paint is recycled.

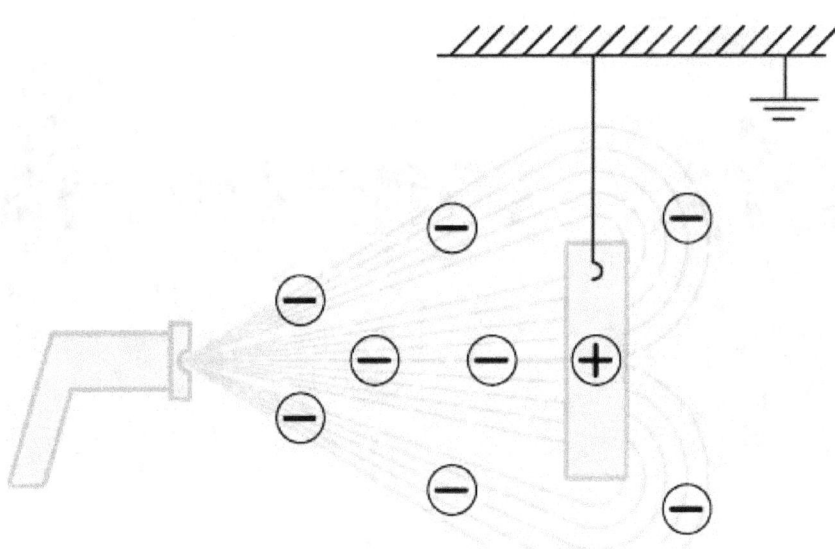

The "-" paint is
attracted to the "+" frame.

Colors make cars look nice. Paint is also a protective coating for cars. It keeps iron in the sheet metal from rusting.

Next, we need somewhere to sit inside the car.

Seats

Seats are made this way. First, weave strong fibers into cloth fabrics. Second, cut the colored cloth into pieces.

weave
(close-up)

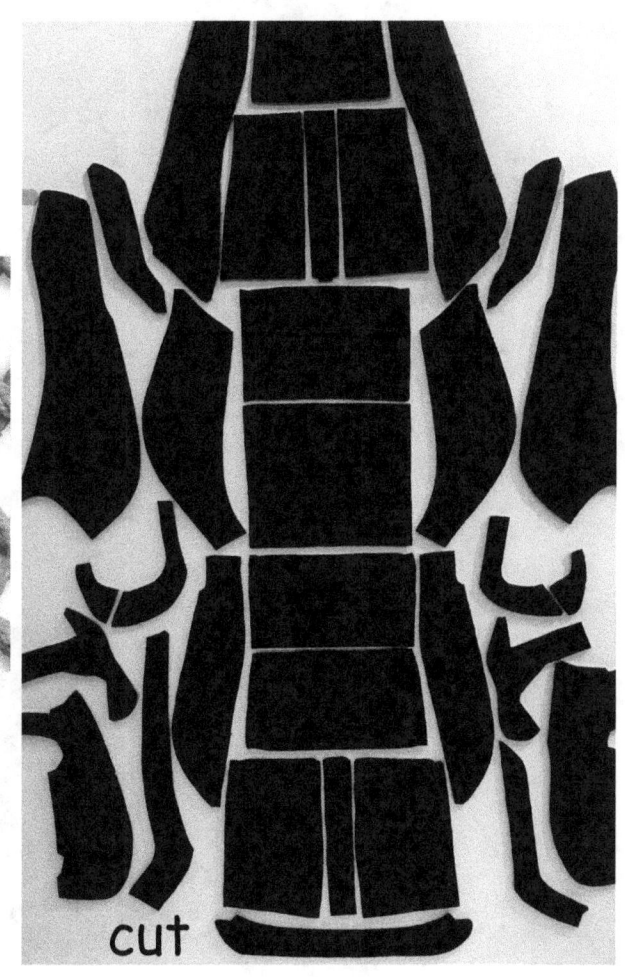

cut

Third, sew the fabric pieces and stuffing together to make high-wear chairs.

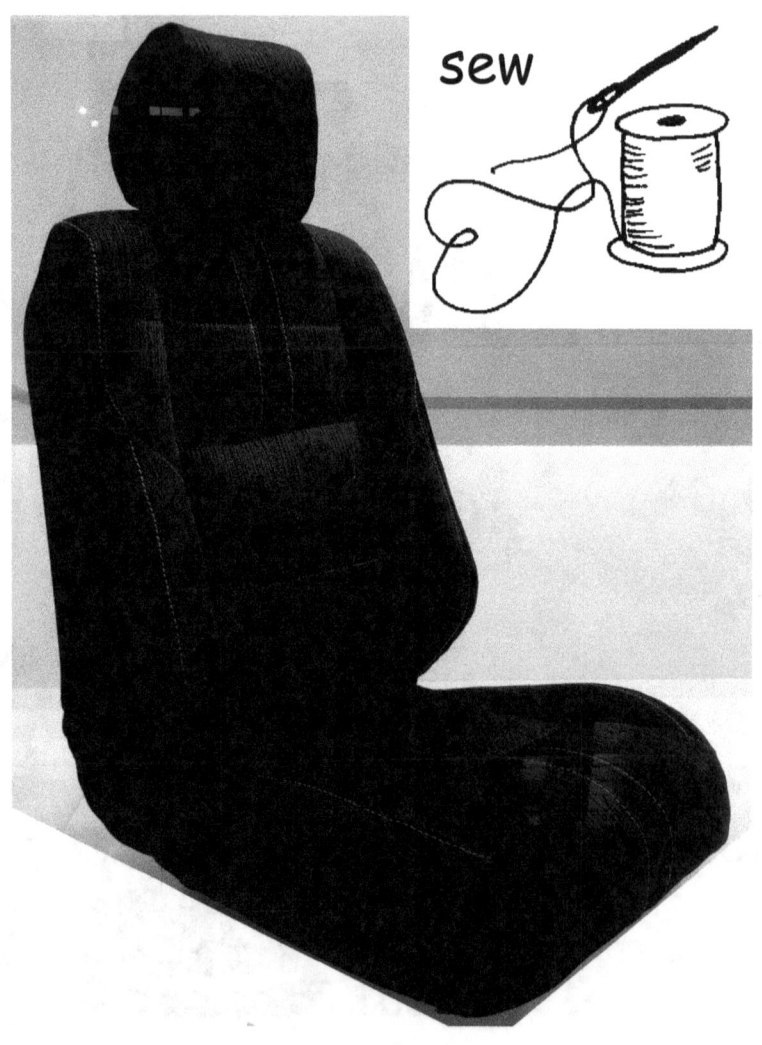

sew

Other Systems

Next, make the other systems. This includes: wiring, cooling and suspension.

Wiring connects the battery to the engine, lights and interior.

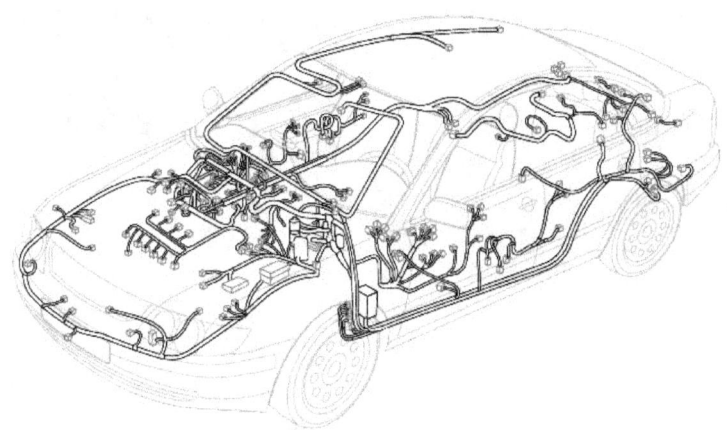

The cooling system keeps the engine from getting too hot.

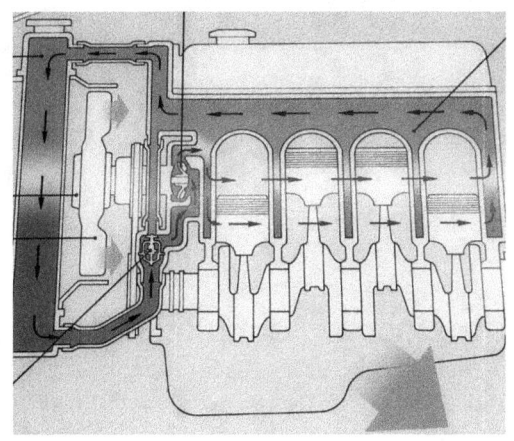

The suspension system keep the ride smooth even over bumpy roads.

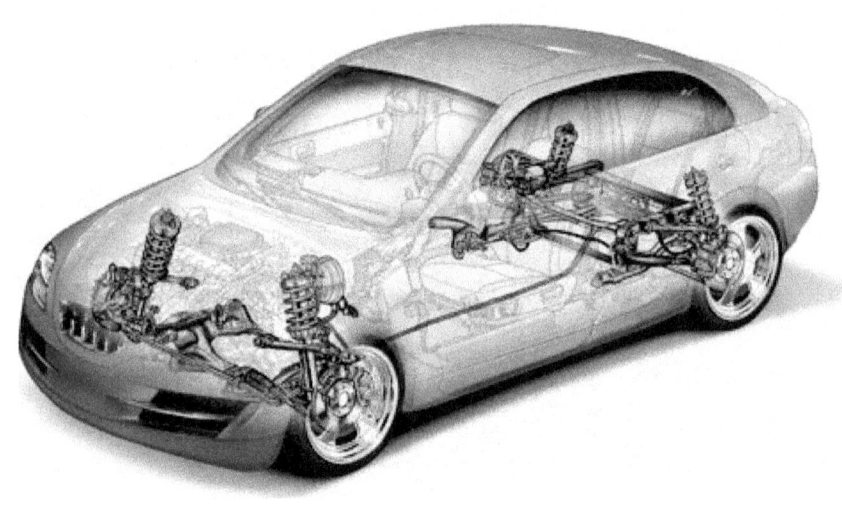

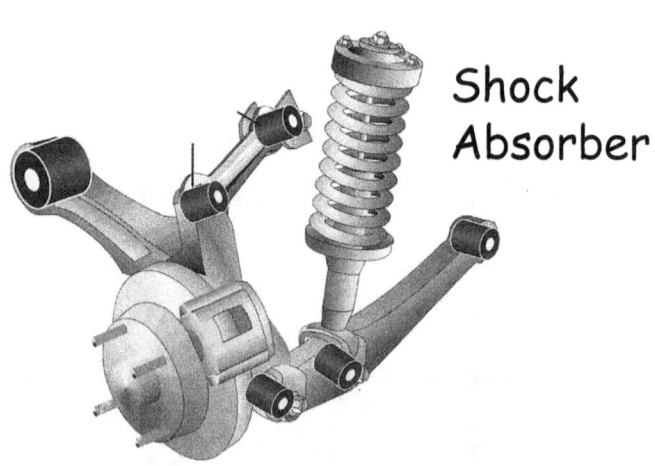

Shock Absorber

Next, join the parts to make the complete car.

7) Put Together

Next, put together or assemble the car using a balance of humans and machines. This automated assembly tool installs the major systems.

Robots

Robots do the heavy
lifting and strong assembling.

People

People do the delicate work
like installing windshields.
The windshield is glued in place.

Torque Turns

Let's look more at the science of
assembly. Parts like the engine,
seats and wheels are joined
together or fastened with
nuts and bolts.
Twisting together the nuts and bolts
is called "torque." Tight torque is
what keeps the parts together.

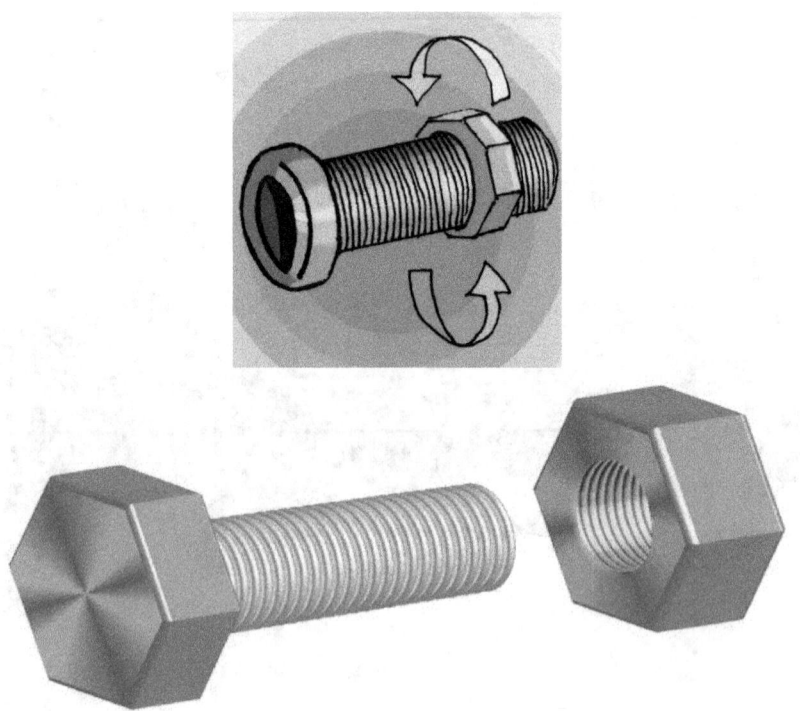

More Torque

Nuts and bolts join parts together. They are stronger than spot welds.

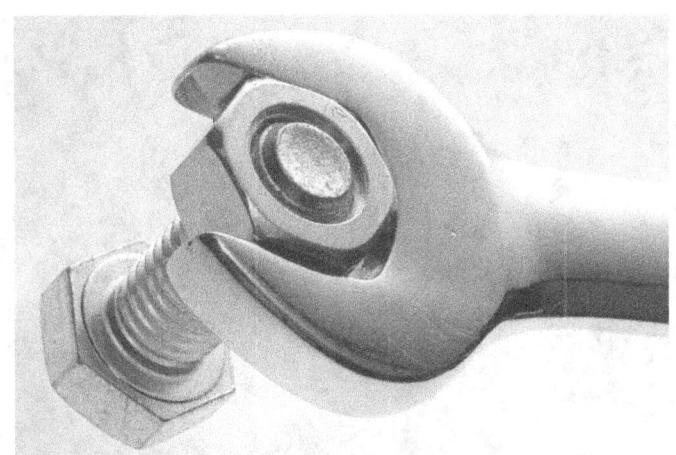

Nuts and bolts are also used to join parts that will need to be taken on and off again like wheels.

Pass Test

The car is tested to make sure
all the systems work properly.

Done

Now, the car is complete!
It is ready for us to buy
and drive!

Conclusion

Science is in every step to make cars.
Pour the engine with heat. Pound
crankshafts until groups of atoms
line up. Cut apart gears with triangle
tools. Push sheet metal into molds
with pressure to form frame parts.
Spot weld the frame assembly with
many small circles of electric heat.
Paint with charged spray. Finally,
put together the car with twisting
torque. Simply said, the science
of parts, processes, machines
and people make cars.

For most of our history,
we humans walk, ride on
horses or drive horse pulled
carts to get around on land.
With science, separate parts
combine to make a self-
moving car called an "automobile."
Cars give us the freedom to go
anywhere there are roads and
maybe one day, places without
roads — the sky!.

Cars

Cars!

Credits

Page   **Description**
Front Cover, v, 7,8, 12 (Red), 14, 19(Left), 25
(Scissors), 30, 33, 34, 35, 36, 39, 44, 45, 55, 56,
57(Top), 60, Drawn by Joshua C. Bugayong. Property
of Alford Books.

Page   **Description**
12   Free Use. www.robotics.org

Page   **Description**
24,   Passakron Samrandee. Property of
                  Alford Books.

Page   **Description**
50 (Weave)   Cozy Clothes. Drawn by Besty S.
                  Hayes. Property of Alford Books.

Page   **Description**
15, 16,18, 20, 22, 27, 37, 38, 40, 41(Frame
Assembly), 43, 46, 47, 50, (Seat) 51, 54, 58
(Lower), 61,   Toyota Museum of Industry
and Technology Nagoya Japan
Note: No endorsement is expressed or implied
by referencing the pictures from the Toyota
Museum of Industry and Technology Nagoya
Japan

All other pictures are Public Domain

Watch VIDEO

Science of Cars!
- 7 MAKE Steps

(CTA-6)

For a copy of this video contact:

When you watch this video, keep in
mind the seven steps to make cars.

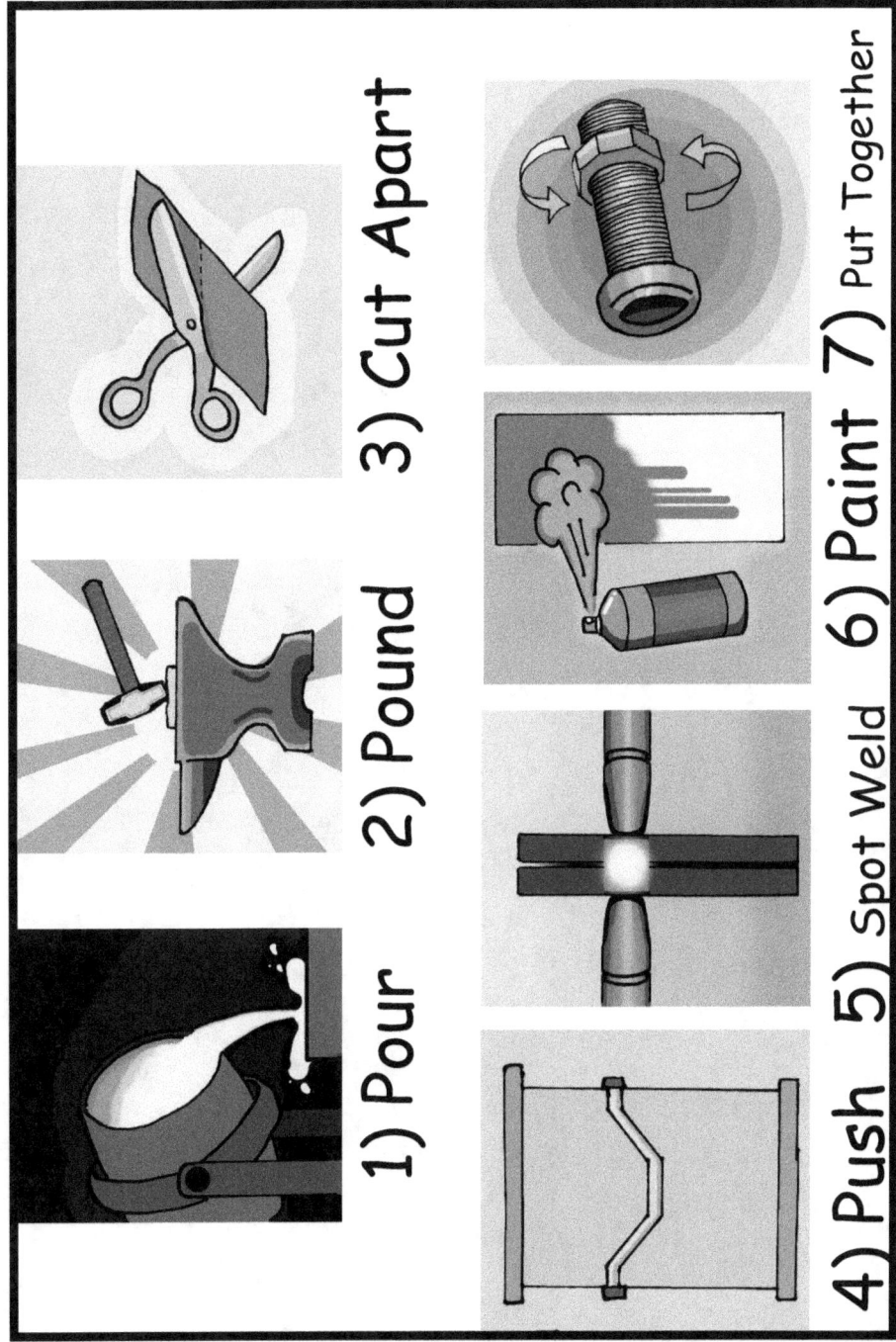

1) Pour   2) Pound   3) Cut Apart

4) Push   5) Spot Weld   6) Paint   7) Put Together

Cars — Main Points

1) Heat melts the metal that
is <u>poured</u> into molds to make engines.

2) Pressure <u>pounds</u> hot
metal to make crankshafts.

3) Wedge-shaped tools
<u>cut apart</u> gears from room temp metal.

4) Pressure <u>pushes</u> sheets of flat, thin metal
between tools to make body frame parts.

5) Electricity heats spots in frame
pieces to <u>spot weld</u> them together.

6) Negatively charged <u>paint</u> is attracted
to the positively charged frame assembly.

7) All the pieces are <u>put together</u>
or assembled to complete the car.

8) Cars go because gas burns in the engine.
This causes heat and pressure. As the gas
burns it puts pressure on pistons that turn
the crankshaft. Gears transmit the
turning to the tires.

This book explains the science of how to make cars. The steps are: pour, pound, cut apart, push form, spot weld, paint and finally put together the parts. See inside, as science and actions are turned into cars.

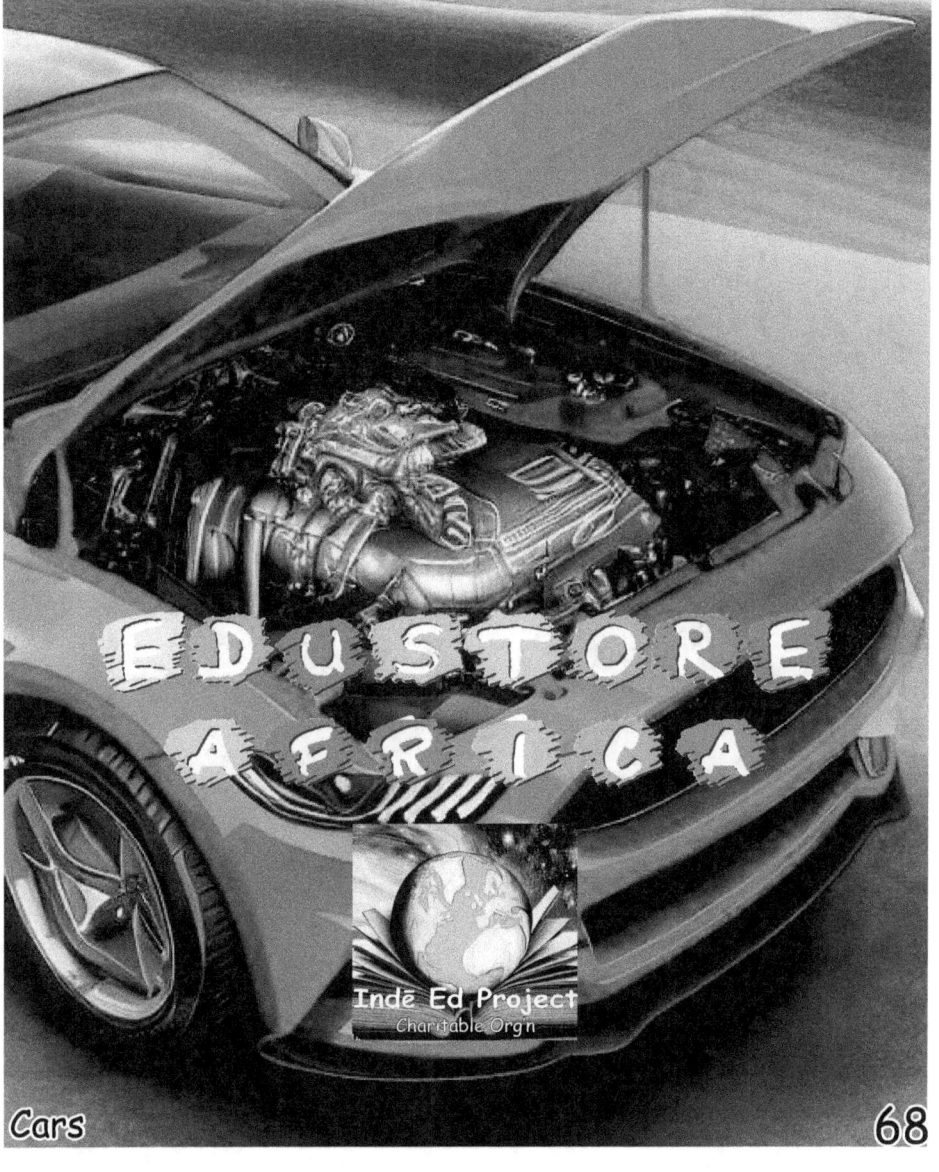

3) Factory
— Clay to Cars

Manu-FACTORY
— Clay to Cars

Science connects factories from ancient clay soldiers to modern self-driving cars.

To manufacture is to make things. Where things are made is called a "manu-factory" or just "factory" for short.

Manu-FACTORY
Clay to Cars

Table of Contents

Without factories,
we produce only a few products.
With science, we build from scarcity to
a surplus of everyday objects for everyone!
Production has challenges like pollution.
Science helps humanity solve problems too.

FACTORY One Pager

Let's see how science connects from clay soldiers to factory cars.

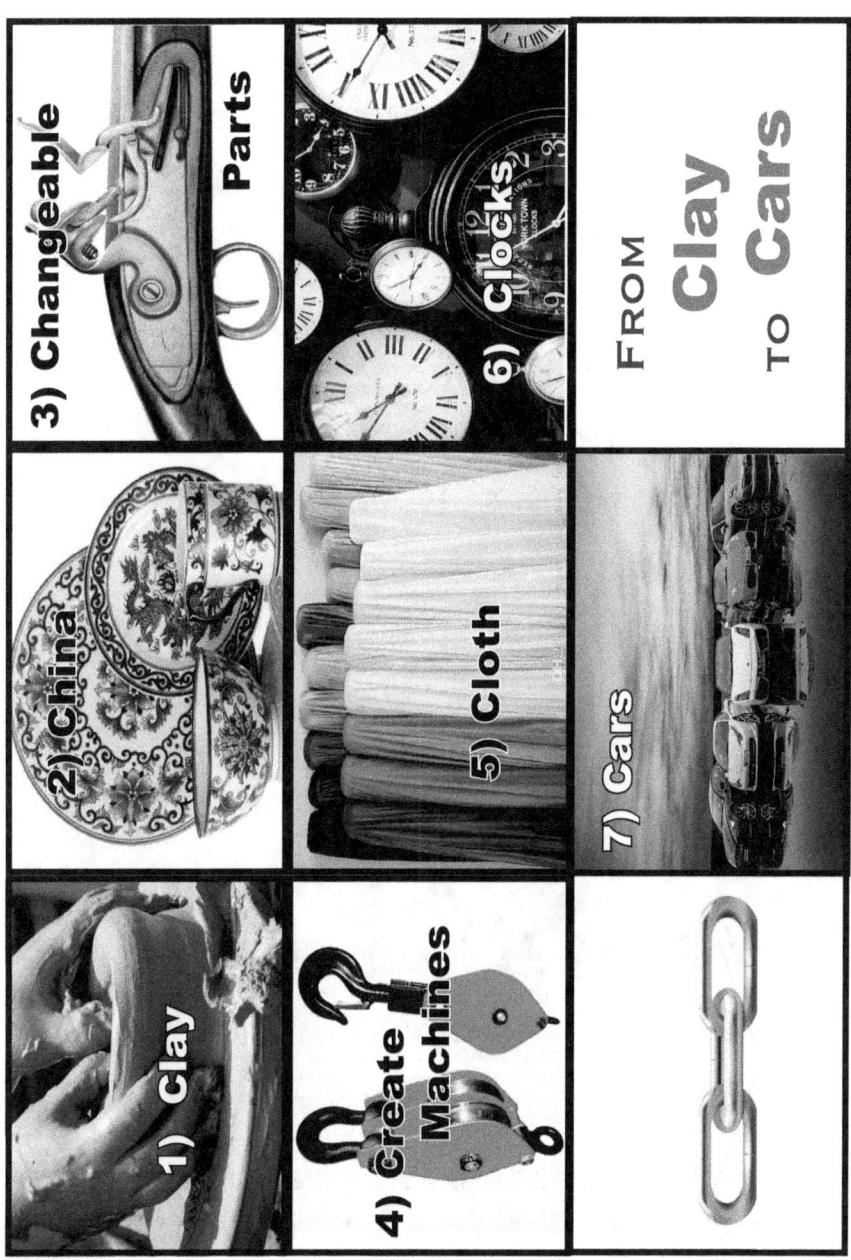

1) Clay
2) China
3) Changeable Parts
4) Create Machines
5) Cloth
6) Clocks
7) Cars
FROM Clay TO Cars

In this chapter, watch for
the following science principles:

- Heat
- Material Science
- Interchangeable Parts
- Mechanical Advantage
- Powered Machine Tools
- Electric Motors
- Machine-made Gears
- Human / Machine Processes
- Assembly Line
- Engrams

People think about fuel, sparks and shafts. Next, our cars move us all over the land and maybe one day soon —the sky!

Present tense verbs are given priority in this book to simplify international versions.

Welcome to the true story with the science of **factory** production!

Science helps humanity go
from a time when most people are
farmers with just enough to eat to a
world full of mass-produced objects.

We look at the origins
of objects to better
understand our lives today.

Do Science!

Our word "tech" comes from the Greek
word téchnē which means "art and skill."
Today, our factories balance human
creativity and craft with the controlled
muscles and movements of machines.

This is the interesting tale
about factories that produce
the products we use daily.

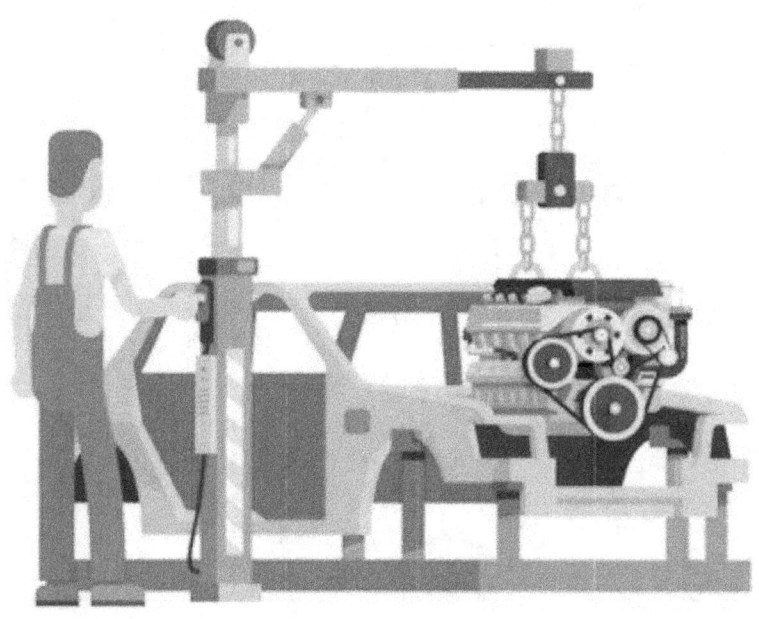

Technology applies science to
produce practical, useful objects.

Also, our tech connects through time from ancient artifacts to modern objects.

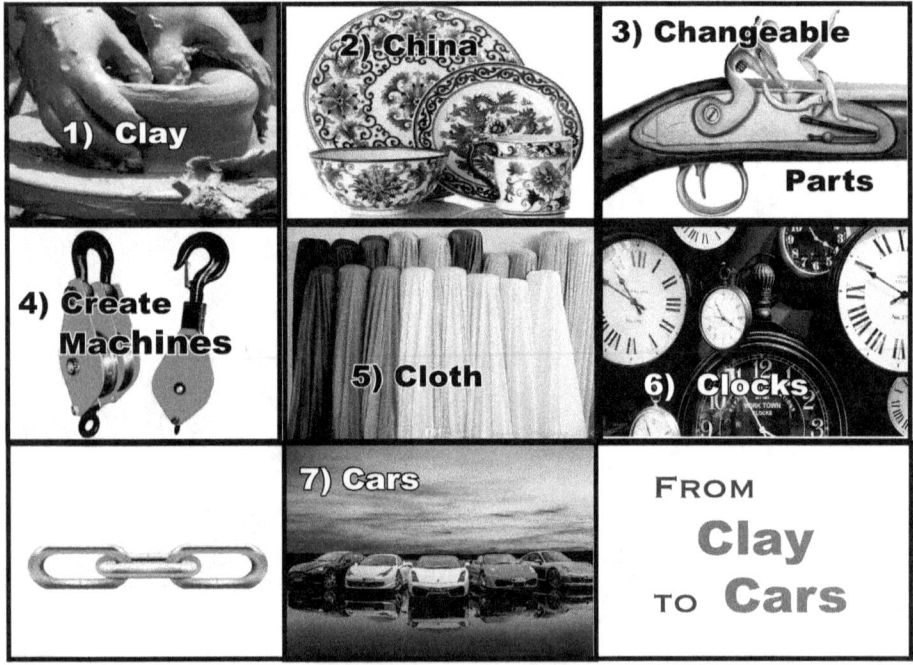

1) Clay
2) China
3) Changeable Parts
4) Create Machines
5) Cloth
6) Clocks
7) Cars
FROM Clay TO Cars

We often think of factories and mass production as being modern. Actually, our true story starts over two thousand years ago.

In the year 221 BCE,
China is a bunch of different
warring kingdoms. Each separate
part has its own war-lord king.

BCE - before common era

Factory

Qin wants to unite the country under one emperor. Himself of course.

Qin Shi Huagdi

He builds factories to hand make lots of weapons like bows and arrows, spears and swords.

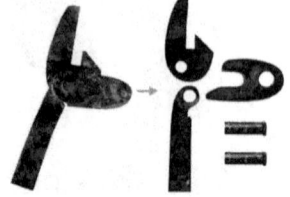

Bronze crossbow triggers are mass-produced in ceramic molds.

Factory

His factories make enough
weapons to equip his 100
thousand soldiers.   Over the next
20 years, Emperor Qin (Chin)
conquers the whole country.
He names it after himself, China.

Factory

ONE, **Clay Army**

He likes being emperor
so much, he wants to keep
the job forever! He has a
plan to rule in the afterlife too.

He builds more factories to hand make lots of clay soldiers and supplies. Over half a million people work 30 years to make Qin's tomb and over 10,000 clay warriors and weapons.

heat

They use some molds but mostly they hand make the clay soldiers.

Think of all the effort and cost to mix the clay, shape it into soldiers, oven-bake the figures and then paint them.

The emperor believes the clay soldiers are real. Somehow the clay army will animate and fight his wars in the afterlife.

The clay army is buried. Two thousand years later, farmers accidentally find them.

As for Emperor Qin, he drinks what
he thinks is an elixir of eternal life.
It contains mercury and slowly kills him.

After Emperor Qin's death, Chinese
factories continue to make silks and ceramics.

Factory

TWO, **Clay China**

China silks and clay ceramics
are popular but expensive. They
are shipped in wooden ships to
other countries around the world.

Factory

Later, people in Europe make
their own copies of the clay ceramics.

Next, in England, Josiah
Wedgwood uses science
to figure out how to make
his own ceramics at less cost.

1760+

Factory

He experiments with what
materials and production
processes to use.
Also, he discovers how to
control oven heat consistently
to make quality ceramics.

Material Science

A fire meter measures temperature
based on how much the clay shrinks.

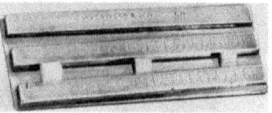

He makes popular ceramic dinnerware and tea sets.

He also makes these fancy products called "Wedgwood."

Through trial and error, Josiah figures
out that the element cobalt makes ceramics blue.

Ceramics are made in factories.

bottle kiln

Today, in the West, we still
call clay ceramics "china"
after where they first come from.

THREE, **Consistent Parts**

In 1785, most things including
guns are still made one part
at a time by human hands.

3-1

In France, this guy gets an idea. Painfully, he has gun parts handmade that are all the same size. This is called "inter-changeable parts."

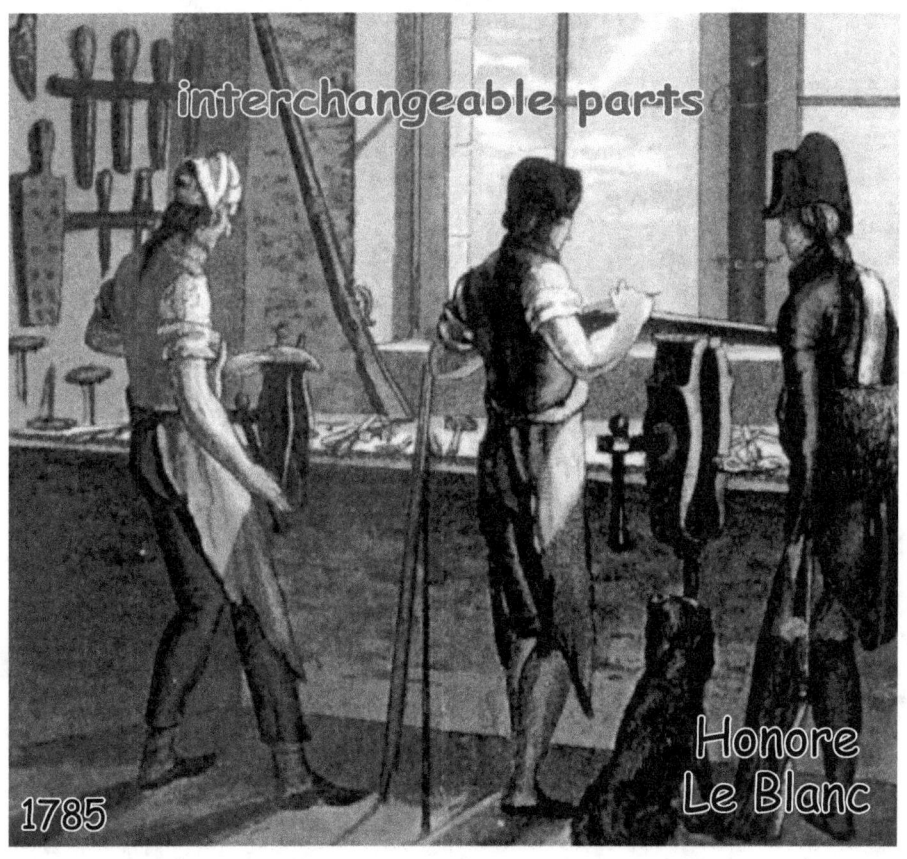

He puts on a big show for important people. In front of the crowd, he puts together random gun parts. The significance being that these parts are interchangeable at a time when other hand made parts are not!

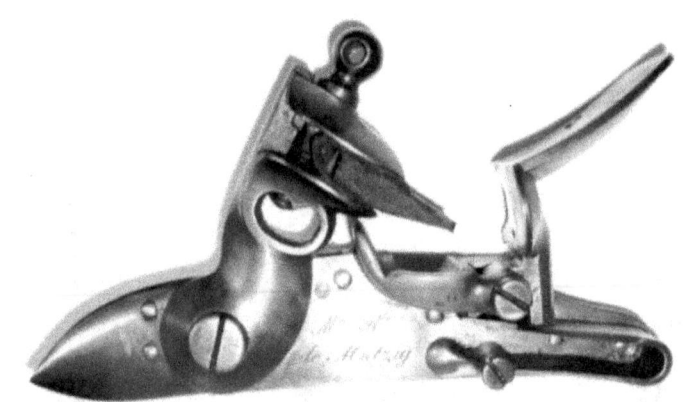

He assembles the "lock" which is the firing mechanism of the gun.

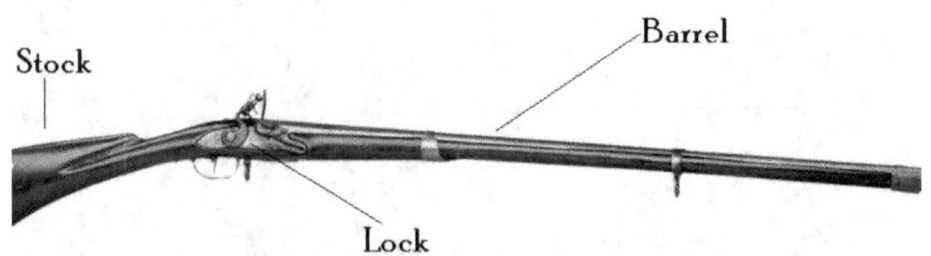

Stock

Barrel

Lock

As a side note, making muskets leads to the English idiom, "lock, stock and barrel" which means "in total or everything."

At least one person in the audience is impressed. He is the US Ambassador to France, named Thomas Jefferson.

Suddenly, in France, a revolution happens. Society unravels into chaos. Heads roll.

1789

The French lose interest in the idea of interchangeable parts.

3-5

FOUR, **Create Machines**

Mister Jefferson, who later becomes President Jefferson, takes the idea of interchangeable parts to the USA.

Eli Whitney and his team design
new water-powered machines
to help make muskets.
It takes 10 years to make
10,000 muskets with
some interchangeable parts.

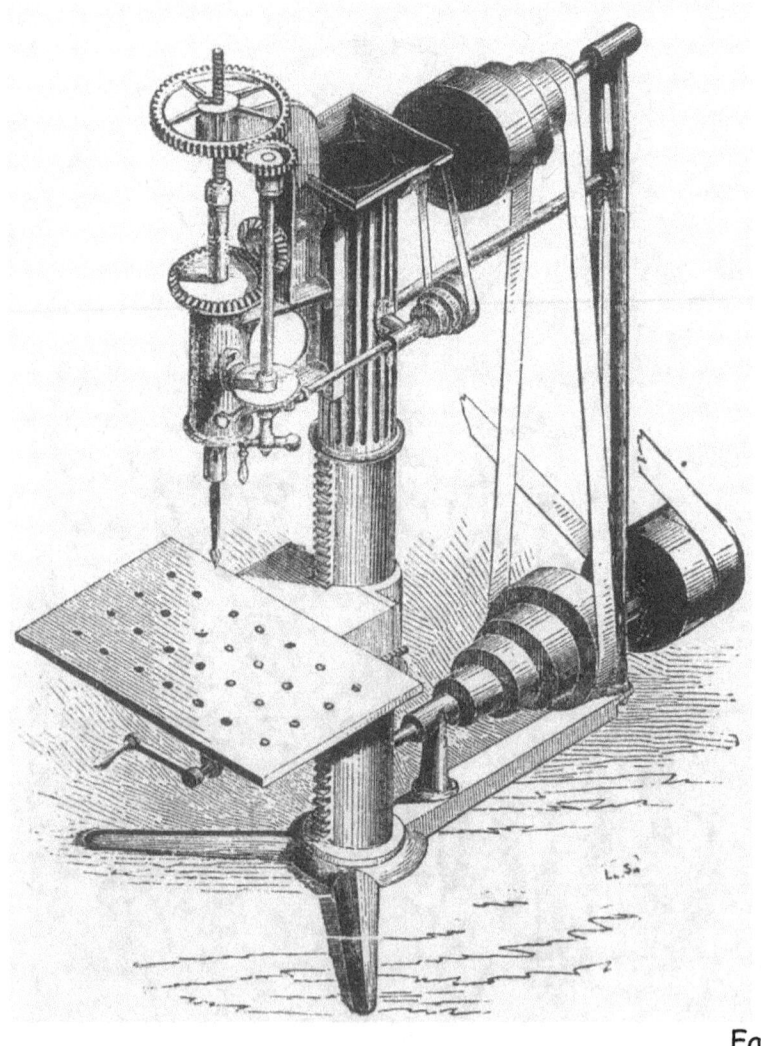

In 1808, England is making
lots of wooden warships preparing
for war with French Napoleon.

1808

There is a problem. There are <u>not</u> enough pulley blocks to control the cloth wind sails that power the navy ships.

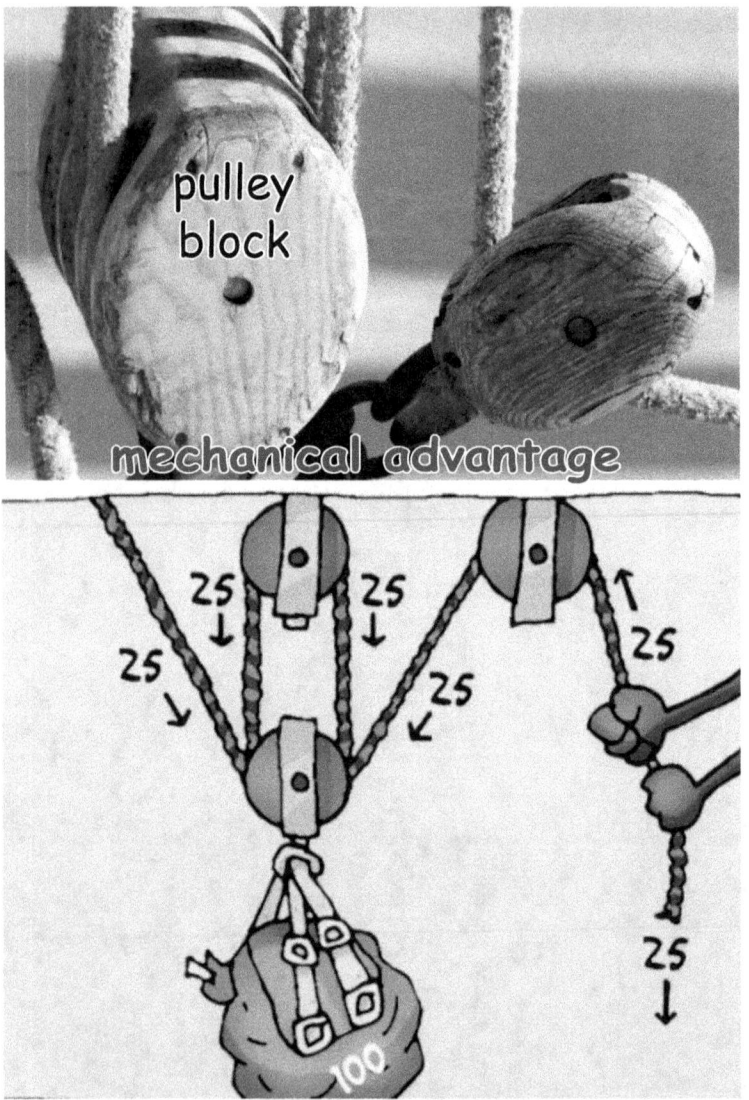

Machines simplify work. Pulleys reduce pull force to 1/4 of the weight but you have to pull the ropes 4 times as long.

Factory

Engineer Marc Brunel gets the idea to create a production line to make lots of wooden pulley blocks. He designs machines but needs someone to make them.

Interestingly, he left France during the French Revolution and moved to England.

Henry Maudslay makes the machines to produce the pulleys. It takes 6 years to make the 45 machines that steam engines power.

Fair Use . Connections by James Burke Pg. 147

England's Portsmouth Block Mill is the world's first steam-powered production line.

Next, the new machines make over 100,000 pulley blocks per year.

Automation impacts people at the pulley factory.
Ten people & machines replace 110 skilled workers.
Britain & allies win the war against Napoleon.

FIVE, Cloth

Next, based on the success
of the pulley block factory,
more machine tools are made.

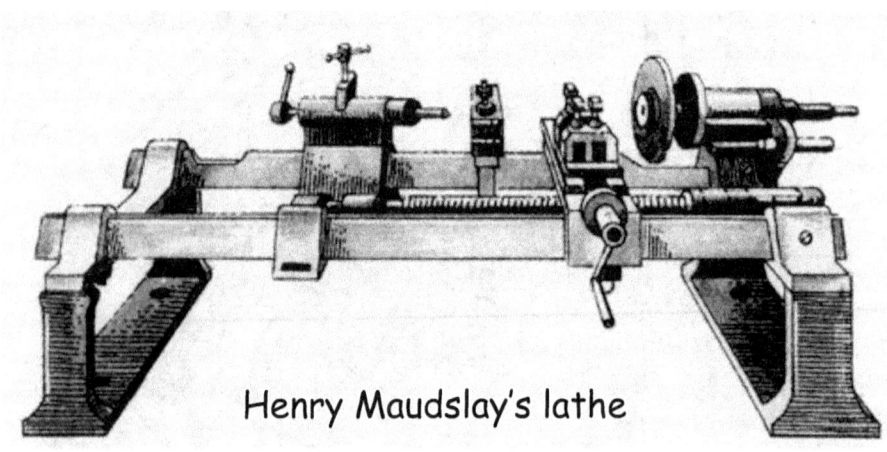

Henry Maudslay's lathe

Eli Whitney's mill

"Machine tools" are machines that make other machines for factories. They also make parts.

powered machine tools

Google search
"Whitworth standard screw threads."

WHITWORTH. & C?
MANCHESTER.
PATENT. N°4. 1842.

James Nasmyth's steam powered drop hammer.

Here are examples of the
factory machines that make cloth.

cloth factory

Machines prepare the cotton.
Carding lines up the fibers.

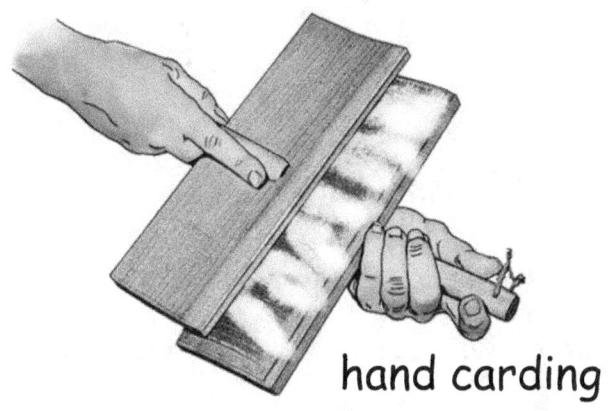

hand carding

machine carding

Other machines spin threads.

hand spinning

machine spinning

Spinning pulls fibers thinner and twists
them together to make threads. Can
we even imagine a time when people, with
much effort, made their own clothes?

Machines like these make products better and more cheaply. But they cost the jobs and change life styles of workers who once made cloth in their homes.

Some workers, called "Luddites" destroy machines. Over time, the machines win.

SIX, **Clocks+**

In the 1820s, this guy has a USA factory that makes hundreds of thousands of clocks using water-powered machines.

Chauncey Jerome

He lowers the price of a brass clock from $20 to $2.

Factory

Over time, water wheels are replaced by steam engines.

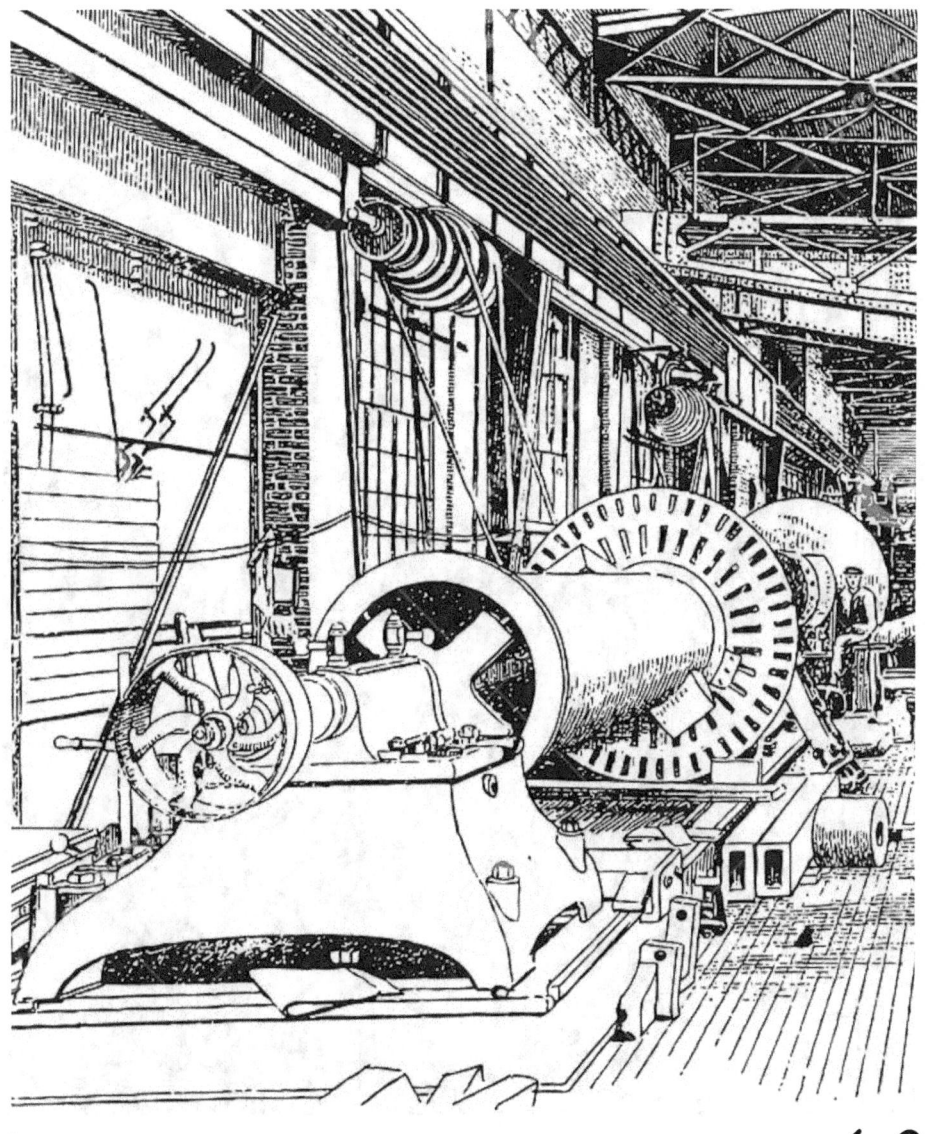

Steam engines power machines in factories. Steam engines turn gears that turn long shafts. Leather belts connect to power individual machines.

The number of
factories increases.
There is an Industrial
Revolution of changes as
workers move from farms
in small rural villages to
factories in crowded cities.

Factory

In the 1850s, this guy combines his own ideas with the patents of others to design a sewing machine. He builds factories and cleverly sells the sewing machines on credit.

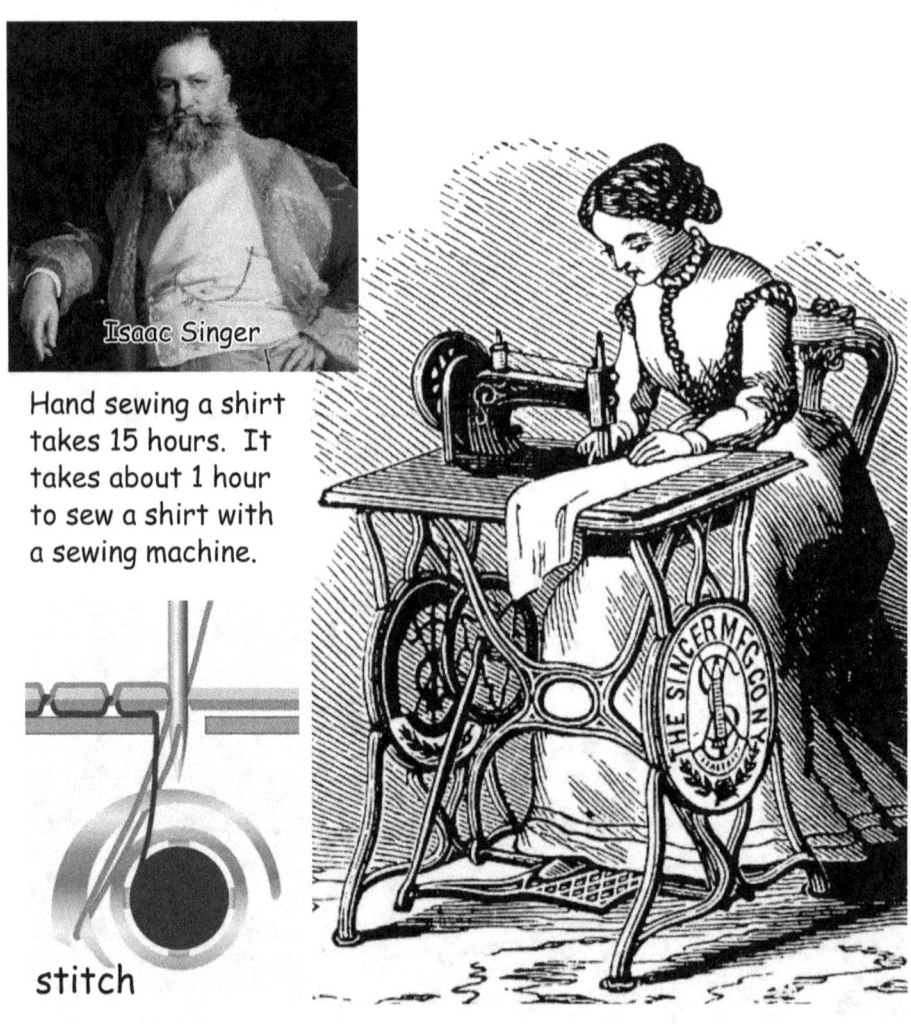

Isaac Singer

Hand sewing a shirt takes 15 hours. It takes about 1 hour to sew a shirt with a sewing machine.

stitch

The needle pushes the 1st thread down that a hook loops around a 2nd thread to make a stitch.

At this time, steam engines power boats and ships that are made in factories.

" The Clermont," Fulton's first American Steamboat.

Notice the smokestack on this river boat. It is similar to the train called "Rocket." Steam engines power both.

Mississippi steamboat

Factory

Later, factories also build equipment to make and distribute electricity.

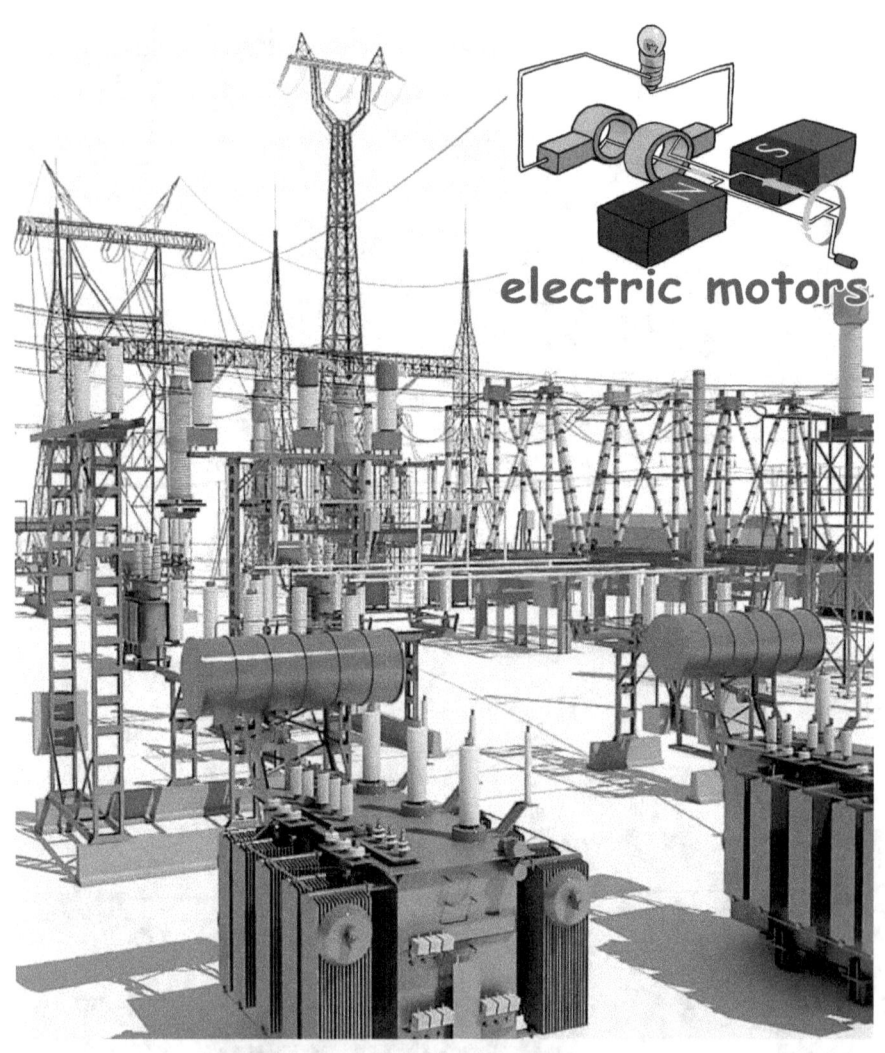

electric motors

Electric motors power more machines and factories.

SEVEN, **Cars**

In 1908, Henry Ford designs a simple but effective car. After many tries, he calls his invention the "Model T."

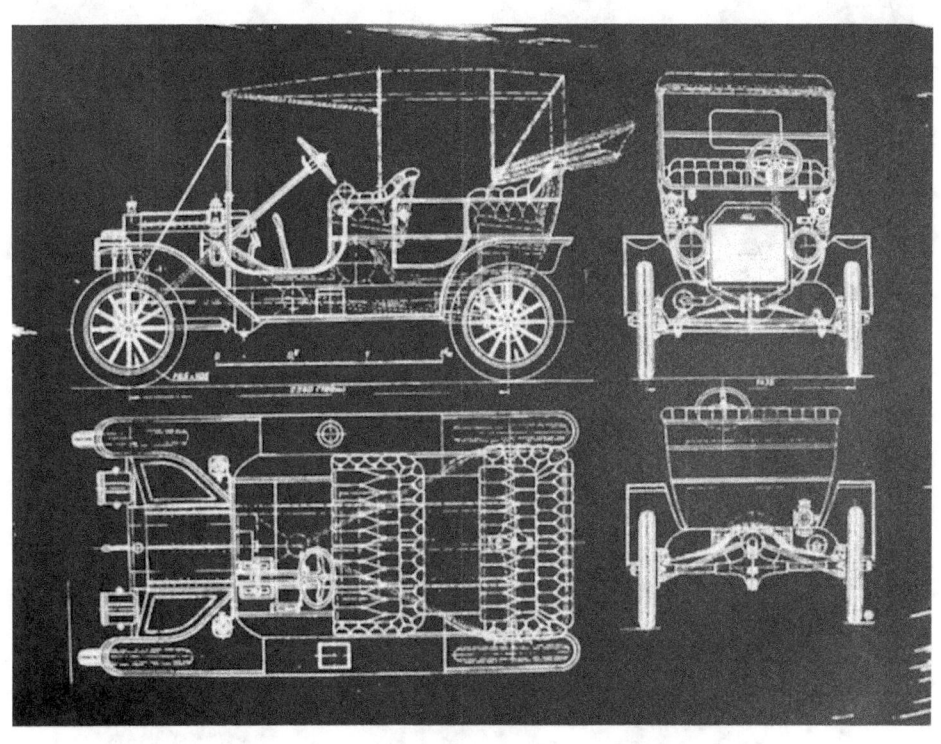

Car pieces are made all the same size. Remember, this is called "interchangeable parts."

7-2

Ford uses machine tools and a moving assembly line to make lots of cars.

He pays his workers a fair wage. Workers can afford to buy the cars they make.

In Ford's factory,
30,000 machines help people
make 15 million cars in 20 years.

Factory

Ford's efficient production line lowers the price of a car from $850 in 1909 to under $300 in 1927.

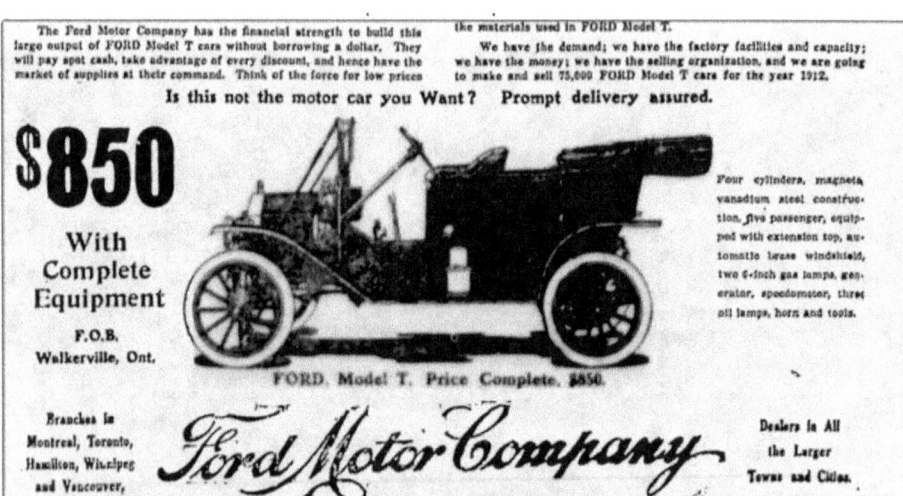

The Ford Motor Company has the financial strength to build this large output of FORD Model T cars without borrowing a dollar. They will pay spot cash, take advantage of every discount, and hence have the market of supplies at their command. Think of the force for low prices

the materials used in FORD Model T.

We have the demand; we have the factory facilities and capacity; we have the money; we have the selling organization, and we are going to make and sell 75,000 FORD Model T cars for the year 1912.

Is this not the motor car you Want? Prompt delivery assured.

$850

With Complete Equipment

F.O.B. Walkerville, Ont.

Four cylinders, magneto, vanadium steel construction, five passenger, equipped with extension top, automatic brass windshield, two 6-inch gas lamps, generator, speedometer, three oil lamps, horn and tools.

FORD, Model T. Price Complete, $850.

Branches in Montreal, Toronto, Hamilton, Winnipeg and Vancouver,

Ford Motor Company

Dealers in All the Larger Towns and Cities.

Ford Touring Car

$295

F. O. B. DETROIT
Starter and Demountable Rims 85g Extra

OF all the times of the year when you need a Ford car, that time is NOW!

Wherever you live—in town or country—owning a Ford car helps you to get the most out of life.

Every day without a Ford means lost hours of healthy motoring pleasure.

The Ford gives you unlimited chance to get away into new surroundings every day—a picnic supper or a cool spin in the evening to enjoy the countryside or a visit with friends.

These advantages make for greater enjoyment of life—bring you rest and relaxation at a cost so low that it will surprise you.

By stimulating good health and efficiency, owning a Ford increases your earning power.

Buy your Ford now or start weekly payments on it.

Factory

Soon there are lots of
factories that make cars
and all kinds of other objects.

Today, factories make almost all of our everyday objects. Paper and plastic products are examples.

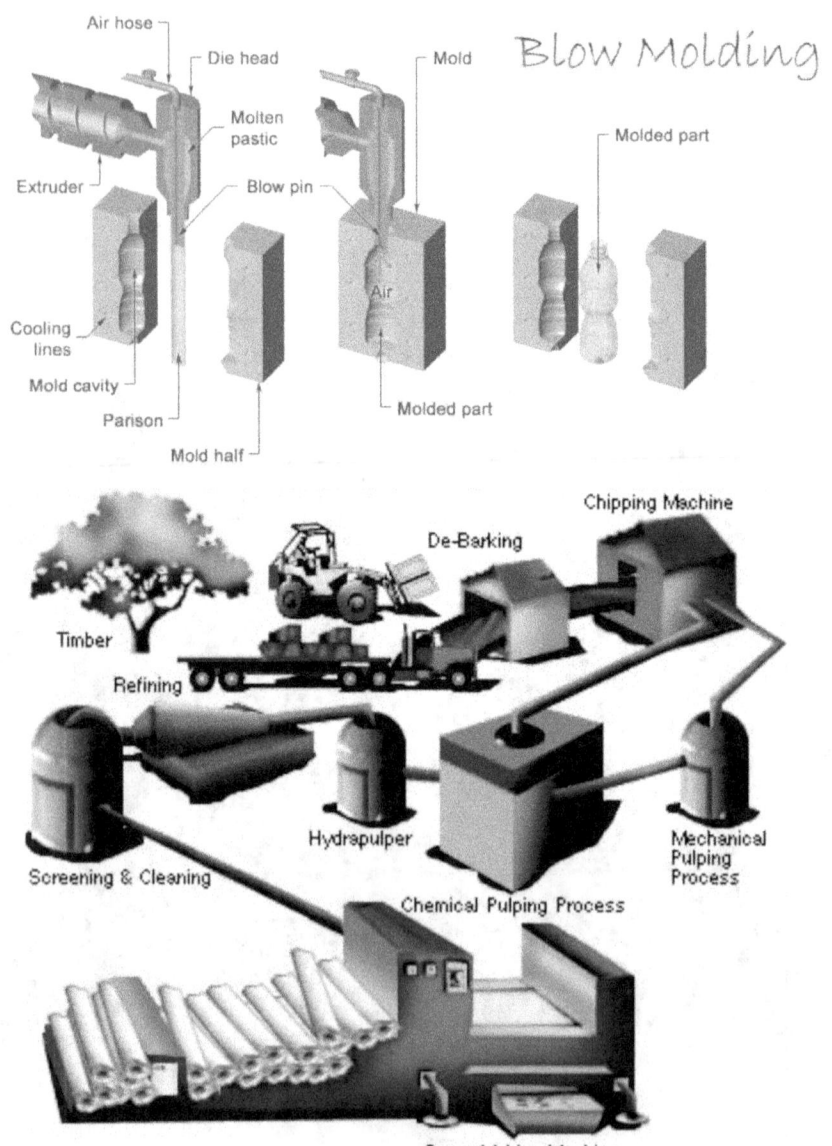

Blow Molding

Air hose
Die head
Mold
Molten pastic
Molded part
Extruder
Blow pin
Cooling lines
Mold cavity
Air
Parison
Molded part
Mold half

Chipping Machine
De-Barking
Timber
Refining
Hydrapulper
Mechanical Pulping Process
Screening & Cleaning
Chemical Pulping Process
Paper Making Machine

Factories shape
and bake clay ceramics too.

Factory

People in factories cut and sew the clothes that we all wear.

Much of what we eat and drink comes from factories.

Factory

Before they make calls and take selfies, smartphones are assembled by people and machines in factories.

8-5

Today, industrial robots,
programmed by people, do many of
the difficult processes to make products.

This example shows factory robots welding a car.

Our everyday objects don't grow in nature. They are made by people and machines. Factories that make microprocessor chips are amazingly complex. Humans design and automate production machines that change sand into silicon chips. These microprocessors empower our everyday digital devices.

clean room

Human brain cells connect together to think. Chip transistors link to enable electronic data to flow. They energize our digital devices today and the emerging AI of tomorrow.

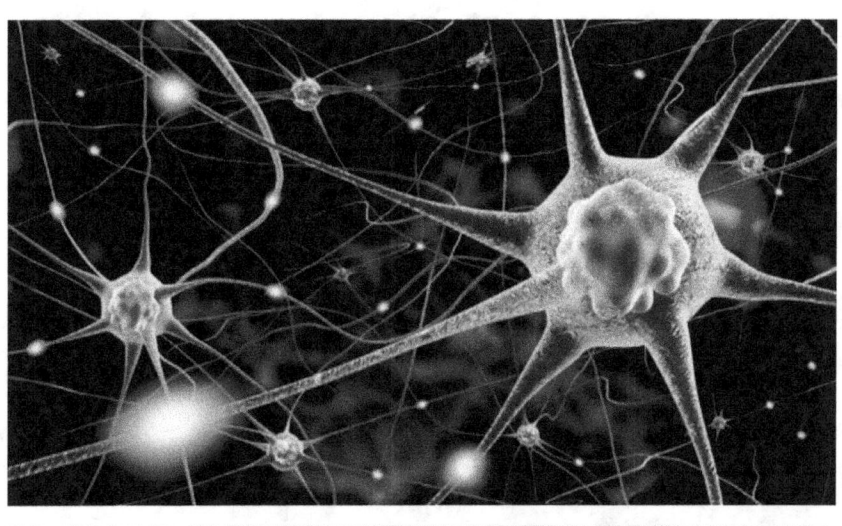

AI = artificial intelligence

To close, Wow! With science,
we see how production factories
connect from clay armies
to colorful cars in 7 steps:

1) Clay
2) China
3) Changeable Parts
4) Create Machines
5) Cloth
6) Clocks
7) Cars

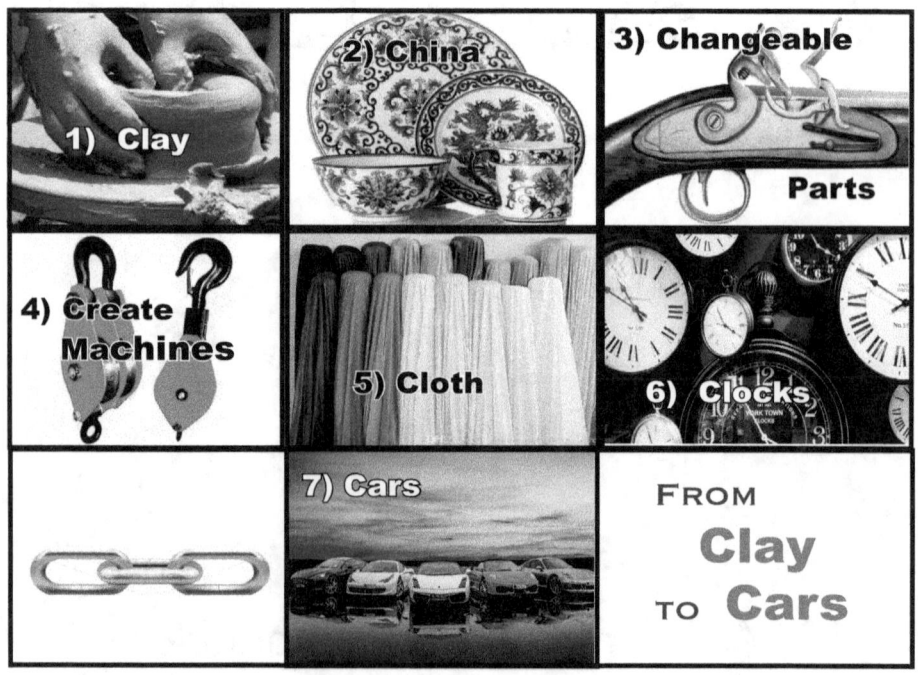

Long ago clay soldiers in China,
show us how early factories
can equip huge ancient armies.

Factory

Next, Chinese factories make lots of ceramics.   They are traded around the world.

Factory

Later, Europeans make ceramics too. New sciences and production processes are discovered.

A guy in France comes up with interchangeable parts. Chaos stops production progress in his country.

Factory

President Jefferson takes the ideas of machine tools and factories to the USA to make muskets with same-size parts.

www.eliwhitney.org

Factory

Then, England invents new
machine tools. First to make
wooden pulleys for ships
and then for more factories.

Fair Use . Connections by James Burke Pg. 147

Next, the powered, controlled machines turn tons of cotton into cloth. The textiles are sold around the world.

Arkwright's water frame

Later, the Ford Factory
uses machines, same-size
parts and a moving assembly
line to make millions of cars.

Factory

Before factories, most people are poor. The rich 1% own the land where the other 99% work. Factories improve economies. Sadly, they create pollution and use up world resources too. Also, many factory workers often dislike the repetitive jobs.

Factory

Today, positively, many people around the world are middle class. But huge education, income and opportunity inequalities still exist. May easy science with the Inde Ed Project help cure the global inequity. We need wisdom as we create our tech-filled and AI-enabled future!

Factory

FACTORY One Pager

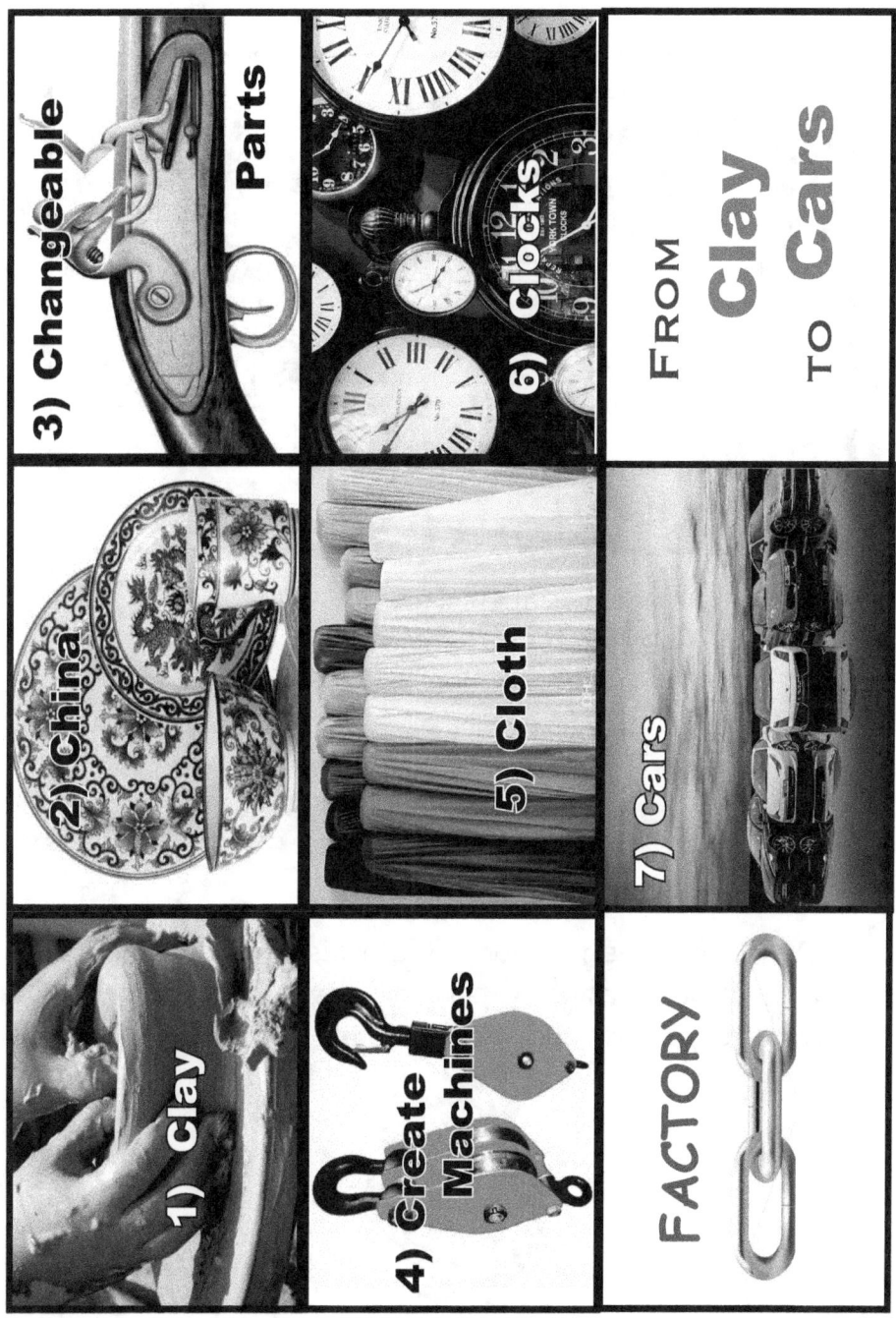

3) Changeable Parts

6) Clocks

FROM Clay TO Cars

2) China

5) Cloth

7) Cars

1) Clay

4) Create Machines

FACTORY

Watch VIDEO

TECH
LINKS
(CTA2-6)
FROM
Clay
TO Cars

For a copy of this video contact:

This is the true tale about factories. Tech connects from ancient armies of clay soldiers to modern cars. Our story starts over two thousand years ago.

3) Factory
— Clay to Cars

Everyday objects made in factories are all around us. Clay is shaped into ceramic cups and bowls. People invent controlled machines to make cloth and clothes. Next, parts are made all the same size that assemble together to make our cars that move us about. Today, humans and machines work together in factories to mass produce our everyday products.

EDUSTORE AFRICA

Indé Ed Project
Charitable Orgn

For Early Learners

Cozy Clozy

— From Fibers to Fabrics!

The science of clothes

<u>Question</u> (Q)
Where do clothes come from?

<u>Answer</u> (A)
Fibers twist into thread
that weave into cloth that
are cut and sewn into clothes.

<u>Key Concepts</u>
. Action steps (processes)
turn separate fibers into
colorful clothes.
. There are different types
of fibers like cotton, silk,
wool and chemical.
. Color can be added to
threads, cloth or clothes.

ii

Word List
fiber

cotton

thread

spin

cloth

weave

color

cut & sew

Action Steps

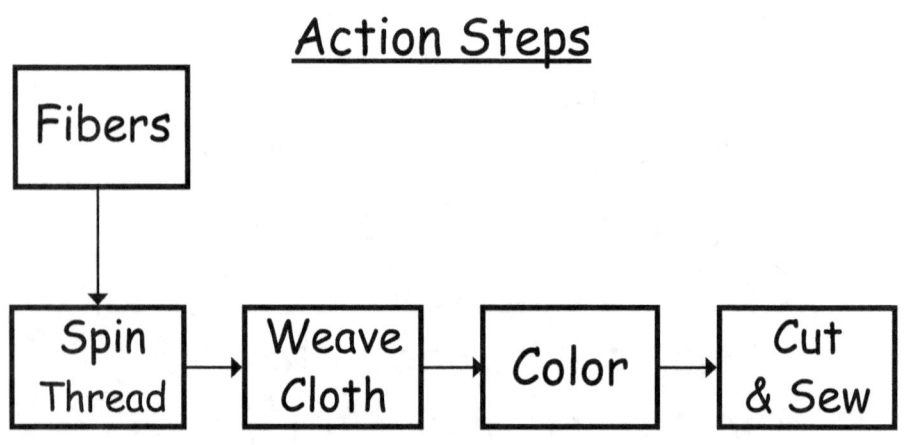

Cozy Clozy

Cozy Clozy

- From Fibers to Fabrics!

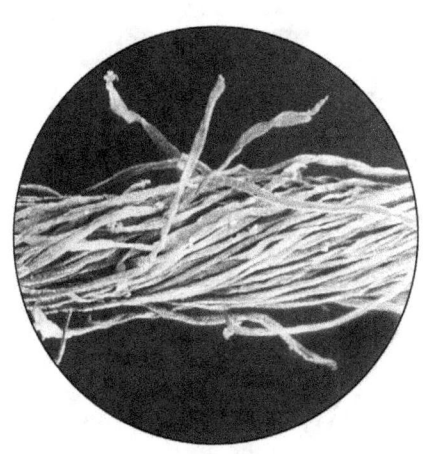

What actions turn cotton into clothes?

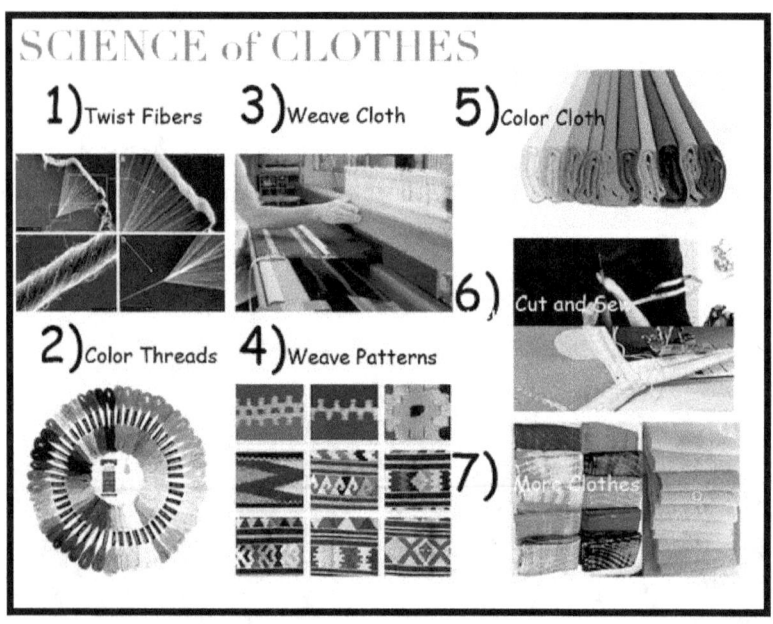

SCIENCE of CLOTHES
1) Twist Fibers   3) Weave Cloth   5) Color Cloth
6) Cut and Sew
2) Color Threads   4) Weave Patterns
7) More Clothes

Hi! I am 'Wormie' the silkworm! Look for me in the story.

The real silkworm is about this size.

v

Cozy Clozy

Table of Contents

Clothes are **cozy!**

Cozy Clozy

Where do **clothes** come from?

shirt

fibers

cloth

thread

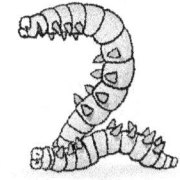

Clothes are made from **cloth**!

Cozy Clozy

Cloth is made from **threads**!

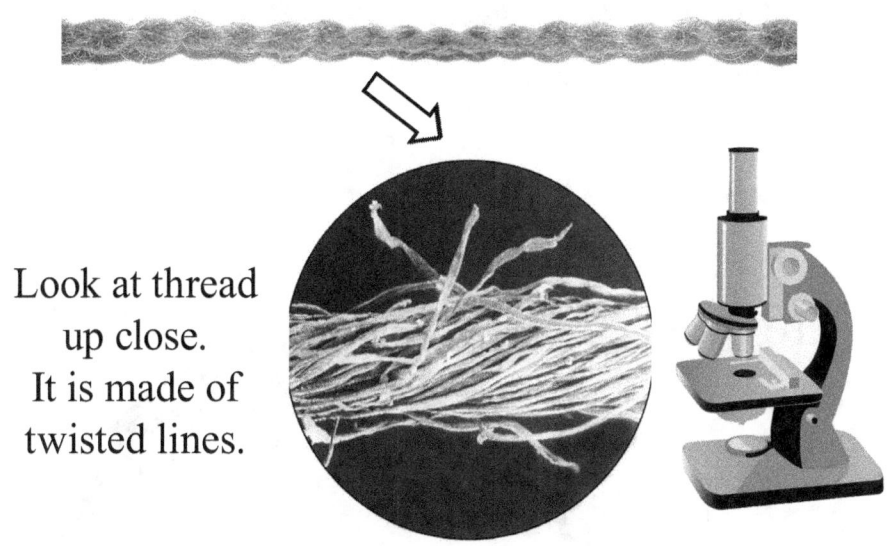

Look at thread
up close.
It is made of
twisted lines.

Do the twist!

Cozy Clozy

Threads are made from **fibers**!

Fibers are thin and long.

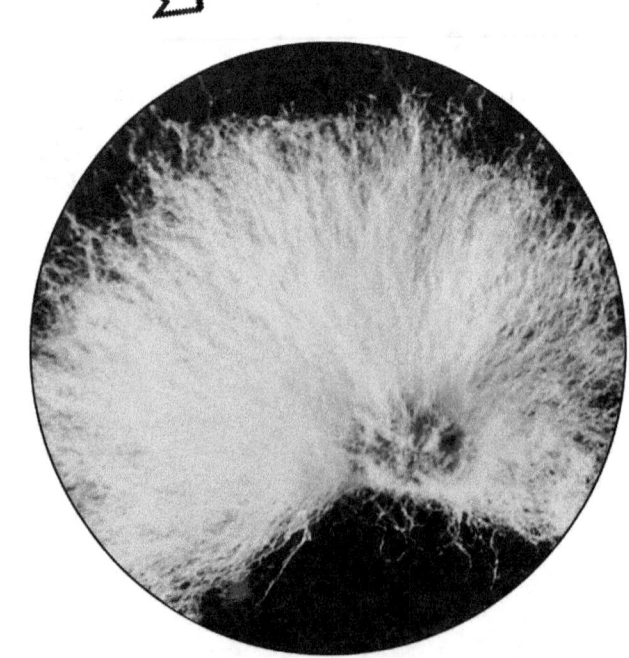

Cotton seed and fibers.

Plants, animals and people make fibers.

People use
chemicals to
make fibers too.

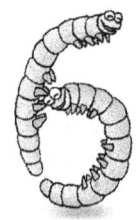

Cozy Clozy

Cotton fibers grow on **plants**.

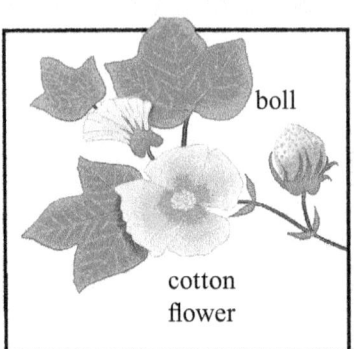

boll

cotton flower

Bugs like the **boll weevil** eat cotton plants.

The real boll weevil is about this size.

It's so soft!

Cozy Clozy

What is made from cotton?
The first airplanes
had cotton cloth wings!

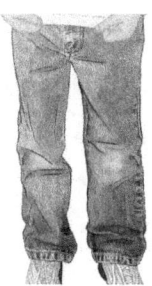  

Today, many clothes are made from cotton.

Cozy Clozy

Did you know that money
is made from the fibers
of two plants?

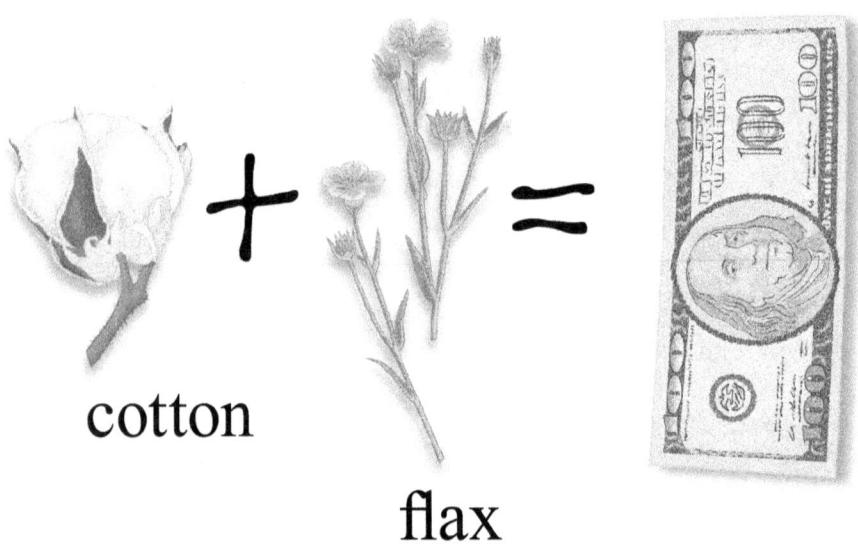

cotton

flax

Flax makes linen!

Wool and silk fibers come from **animals**.

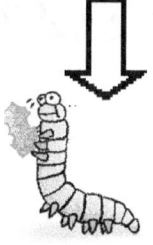

Wool comes from sheep.

Silk comes from what is called a worm but is actually a caterpillar.

Cozy Clozy

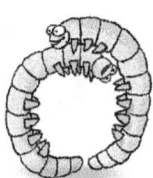

How do you go from moth to cloth?

Yum!

Cozy Clozy

that lays the eggs

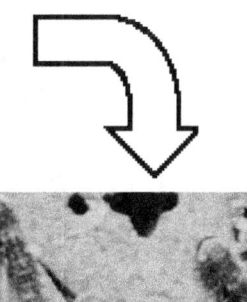

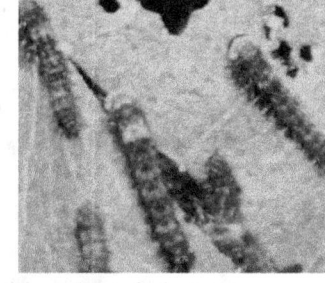

that hatch into worms,
that eat the leaves

This is the moth

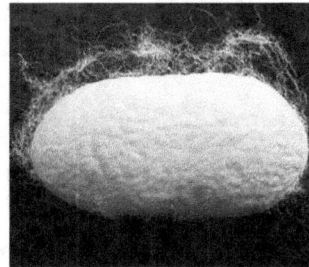

and make silk **cocoons.**

Cozy Clozy

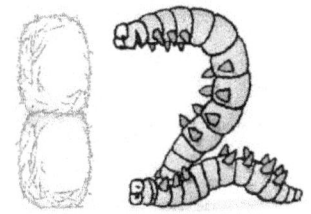

Why is silk
called worm spit?

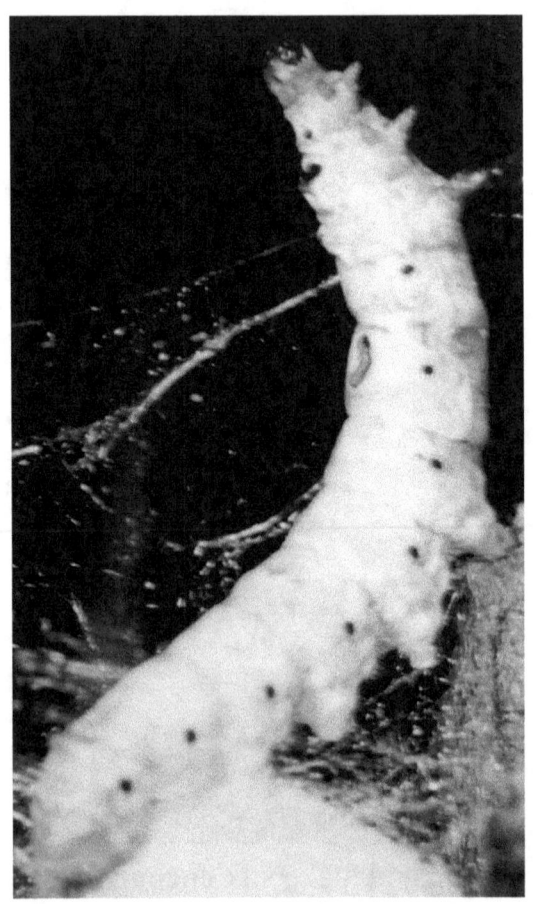

Silk comes out of the
worm's head to make a cocoon.

People make fibers from **chemicals**.

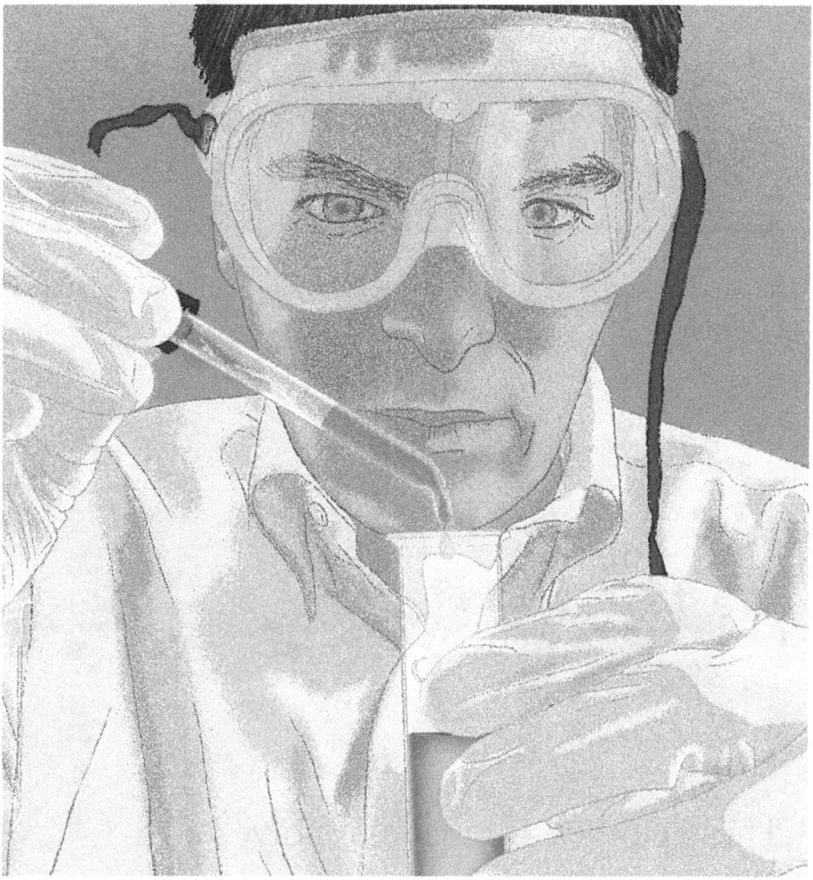

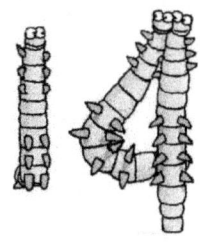

This is how to make **polyester**.

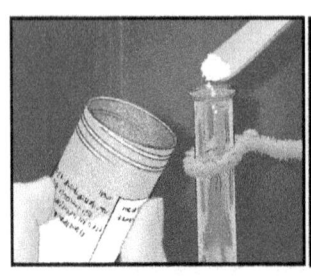 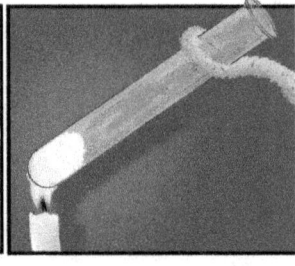 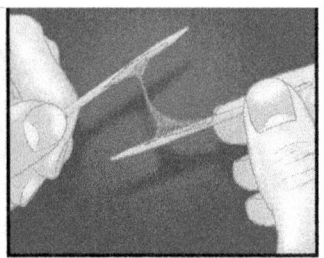

| Mix two chemicals together, | add heat and | pull into thin threads. |

  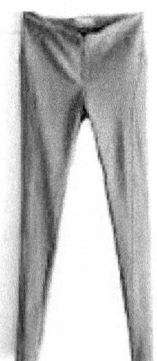

These clothes are made of polyester.

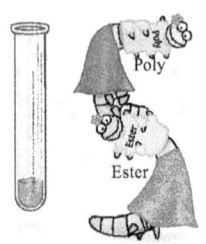

Cozy Clozy

It is interesting that soda
pop bottles with this sign
are made from polyester too.

Fair Use

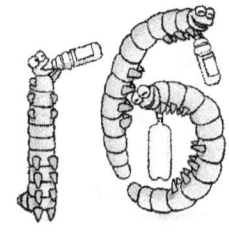

Cozy Clozy

Why do we wear different **fabrics**?

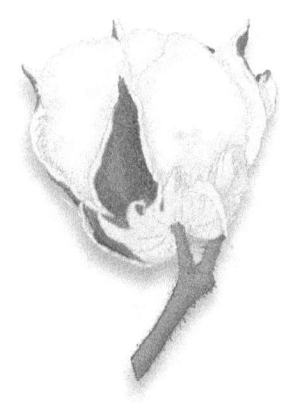

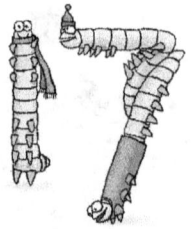

Cozy Clozy

We wear cotton for comfort,   wool for warmth   and silk because it's shiny!

We wear polyester because it resists wrinkles.

Cozy Clozy

Actions turn fibers into cloth!

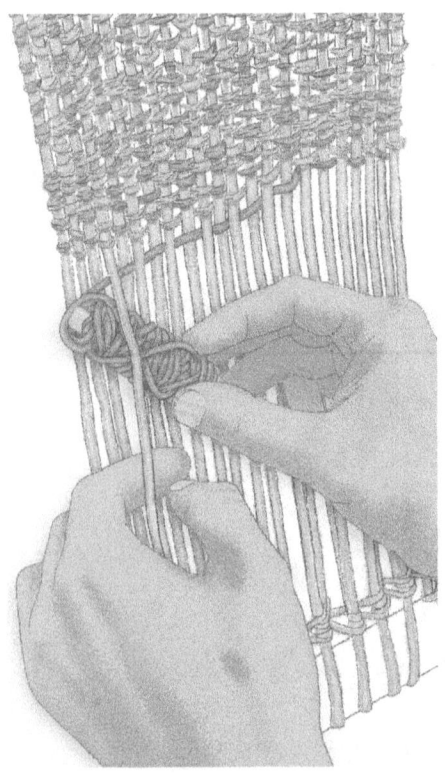

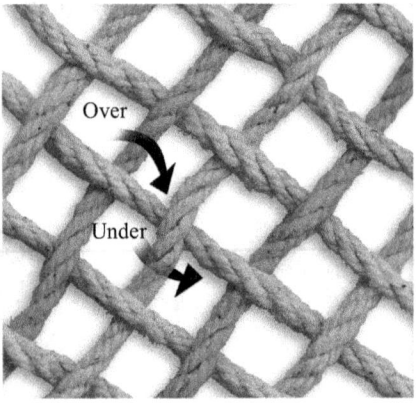

Threads go over and under to weave cloth.

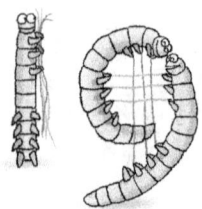

Cozy Clozy

Here are more ways to make **fabrics**:

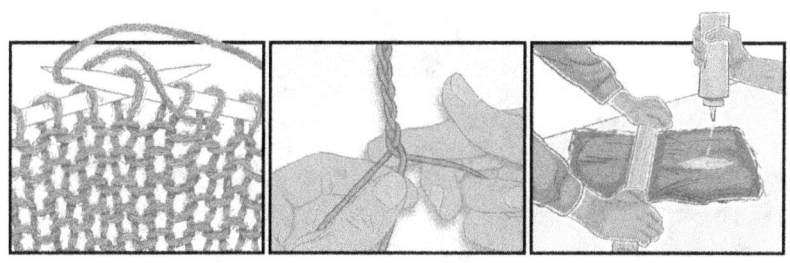

knit loops,   criss-cross **braid**   and smash **felt.**

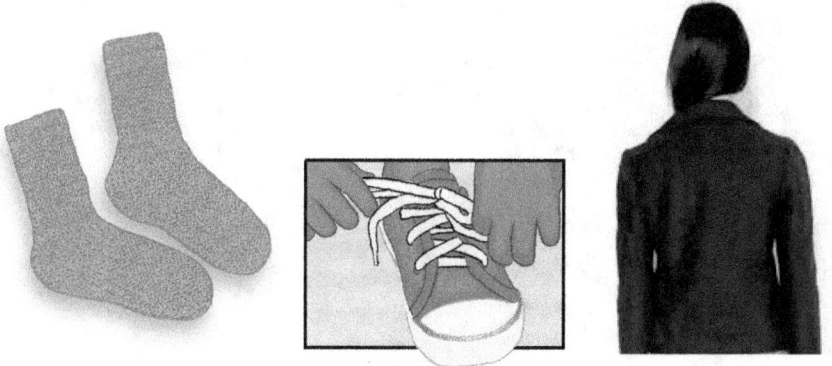

These actions are used to:
knit socks, braid shoe laces and make felt jackets.

Cozy Clozy

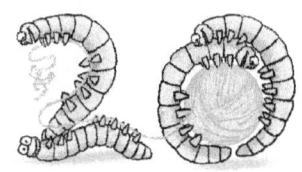

Cloth comes in many colors!

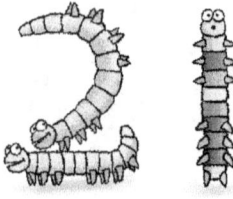

Cozy Clozy

Colors and shapes form **patterns**!

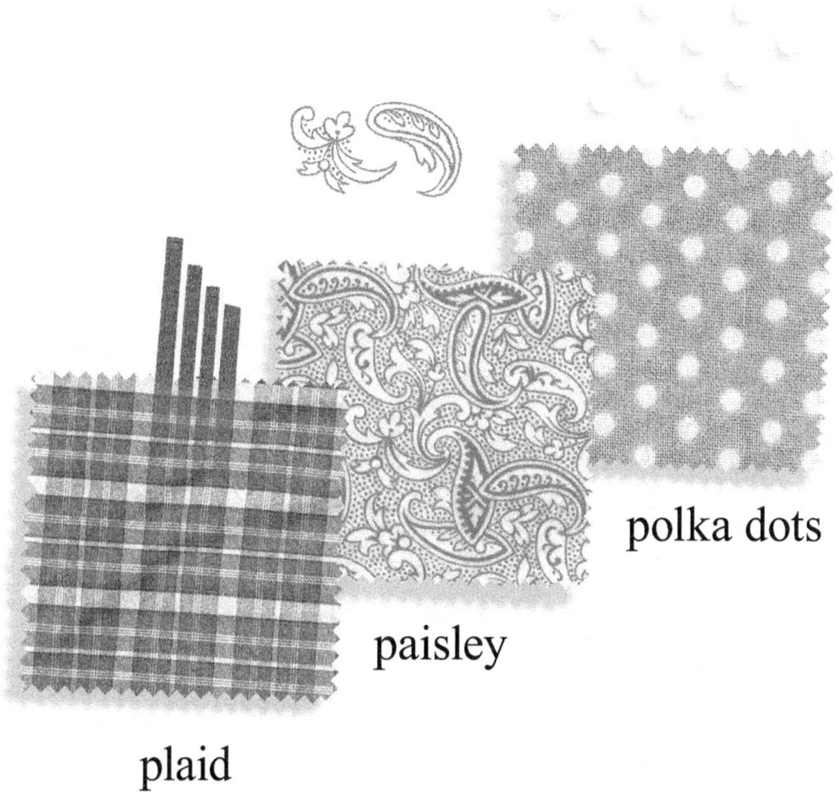

plaid

paisley

polka dots

How are different
patterns put on to cloth?

Tie-Dye

This shirt is **tied** into many parts

to make tie-dye!

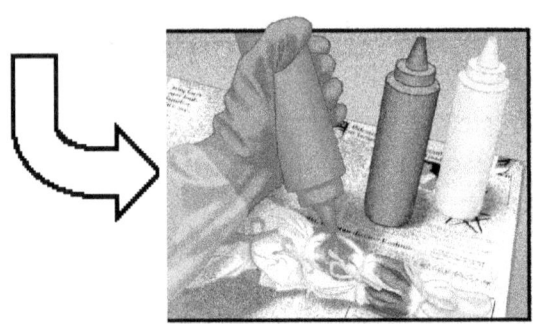

and then **dye** is added

Coooool shirt!

Cozy Clozy

Screen Print

This screen has patterns that are printed onto cloth.

Here is an example.

Cozy Clozy

Roller Print

These rollers print long, repeating patterns. This is the most common way to add color to cloth.

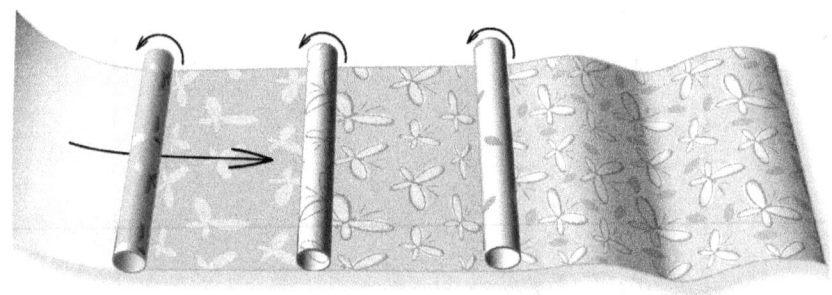

Here is an example.

Cozy Clozy

Cut and Sew

Pieces of cloth are
cut into shapes and
sewn together to
make clothes.

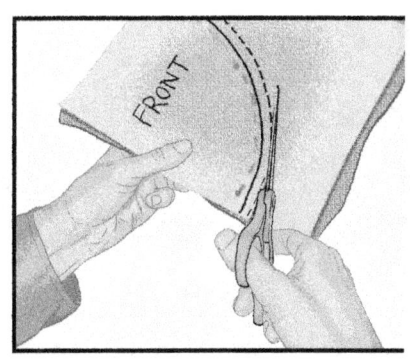

cut

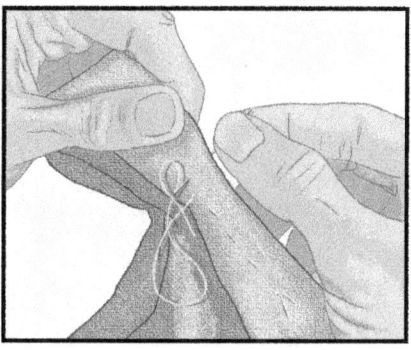

sew

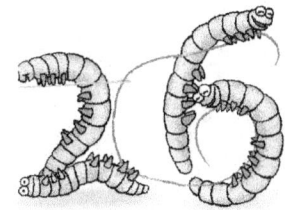

Clothes are the results of these actions:

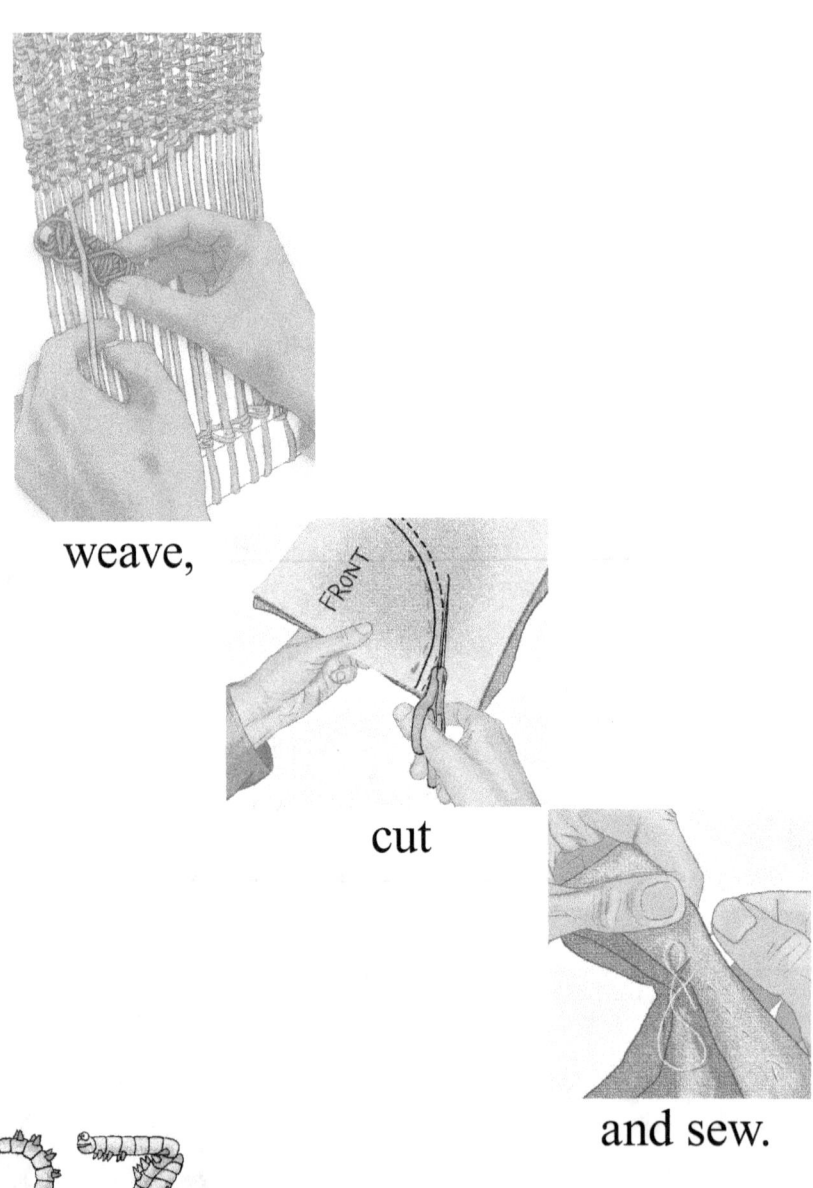

weave,

cut

and sew.

Cozy Clozy

Fibers make more than just clothes.
They also make things like...

curtains,

couches,

and carpets.

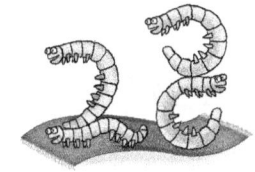

Cozy Clozy

We all use fabrics when we:

work,

play,

leap

and sleep.

Cozy Clozy

Now, to close on clothes.

At first, people
wore animal skins.

Next, people wore
wool and cotton.

Today, we
also wear clothes
made from chemicals.

Cozy Clozy

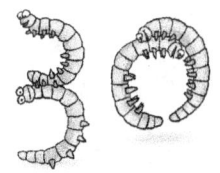

Because of this book, we now
understand how fibers are made
into the fabrics that we use every day!

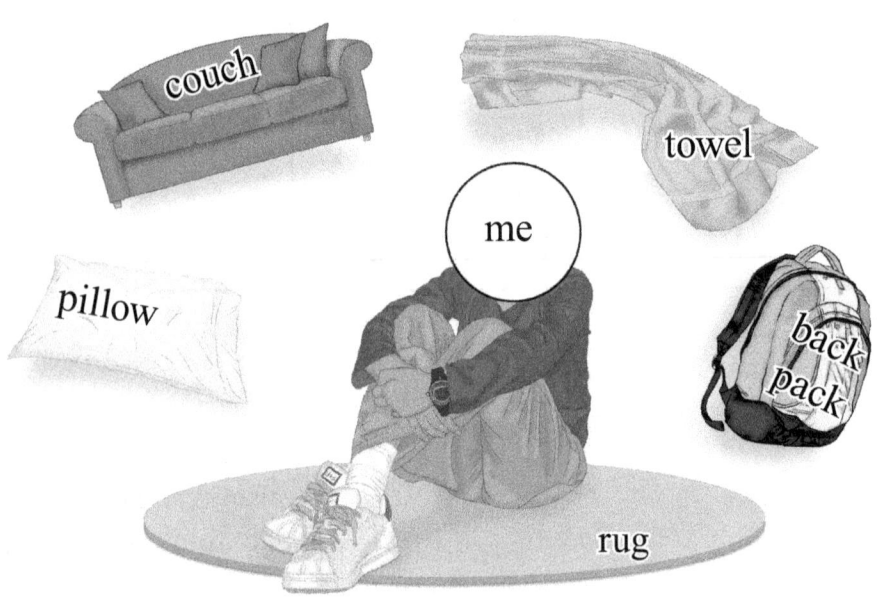

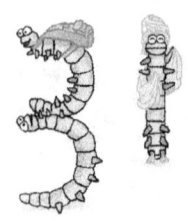

Cozy Clozy

Word List

Word	Definition	Page
braid	Criss-cross threads together.	20
chemicals	Ingredients that mix together to make things	6, 14
cloth	Fabric made from woven threads	3, 19
cocoon	Hard shell made by silk worms. It is made of silk fibers. It's where the caterpillar turns into a moth.	12
cozy clozy	Comfortable clothes	1
fabric	Something made from fibers	20
felt	Glue and smash threads together	20
fiber	Something thin and long that is used to make threads	5
knit	Loop threads together	20
pattern	Colors and shapes	22
polyester	A fiber made by people from chemicals.	15,16
roller print	Apply color to cloth using sequential rollers. This is the most common way to add patterns to cloth.	25
screen print	Use a screen (similar to a window screen) to squeegee color onto cloth.	24
sew	Join pieces of cloth together with a needle and thread	26, 27
threads	Fibers twisted together into thin and long strands	4, 5
tie-dye	Tie cloth into parts and dye to make patterns.	23
weave	Move threads over and under each other to make cloth.	19, 27

Credits

Thanks to the following
for their help with this book!

Quarry Bank Mill
Cheshire, United Kingdom
http://www.quarrybankmill.org.uk

Science and Industry Museum
in Manchester
Castlefield, United Kingdom
http://www.scienceandindustrymuseum.org.u

National Cotton Council of America
Memphis, TN USA
http://www.cotton.org

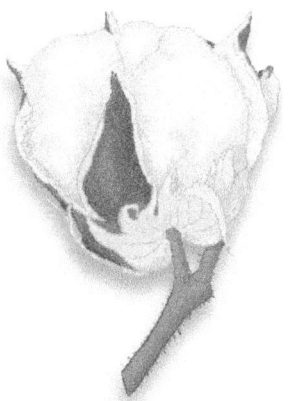

Try It!

Look at labels to see what fabrics are made from! What actions add colors? How are the fibers made into fabrics?

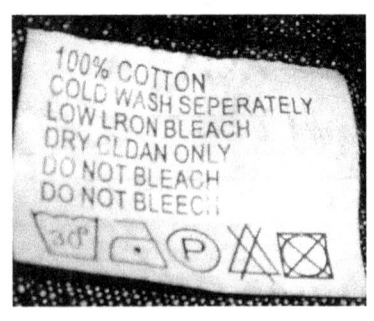

100% PURE WOOL
HAND WASH
LAY FLAT TO DRY

50% COTTON
50% POLYESTER
MACHINE WASH WARM
TUMBLE DRY LOW
REMOVE PROMPTLY
DO NOT BLEACH
MADE IN U.S.A.

100%
SILK
DRY CLEAN
ONLY

Review

From fibers to fabrics, it takes many actions to make our closets full of colorful, comfy clothes.

fibers

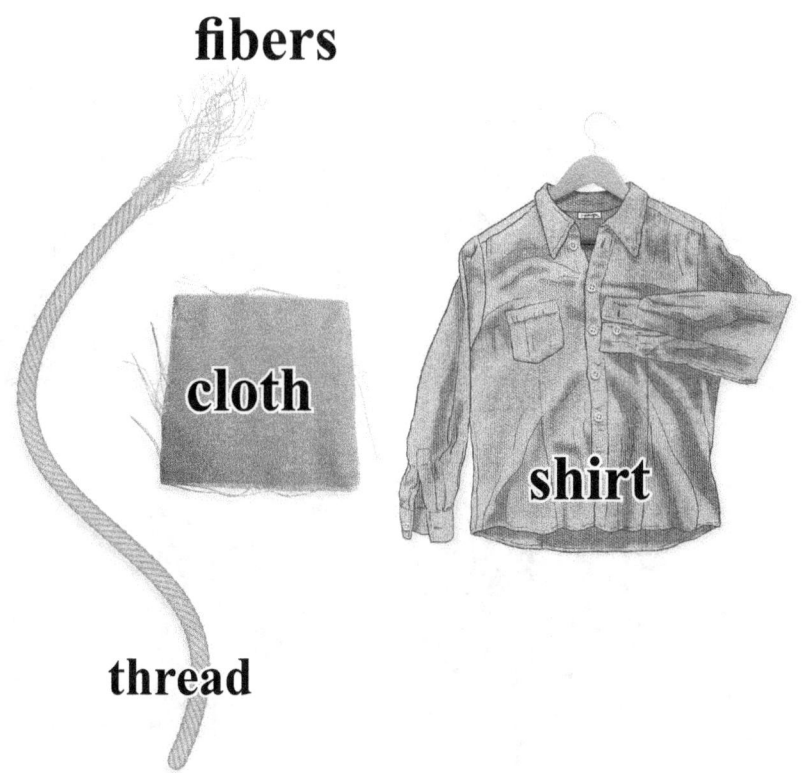

cloth

shirt

thread

SCIENCE of CLOTHES

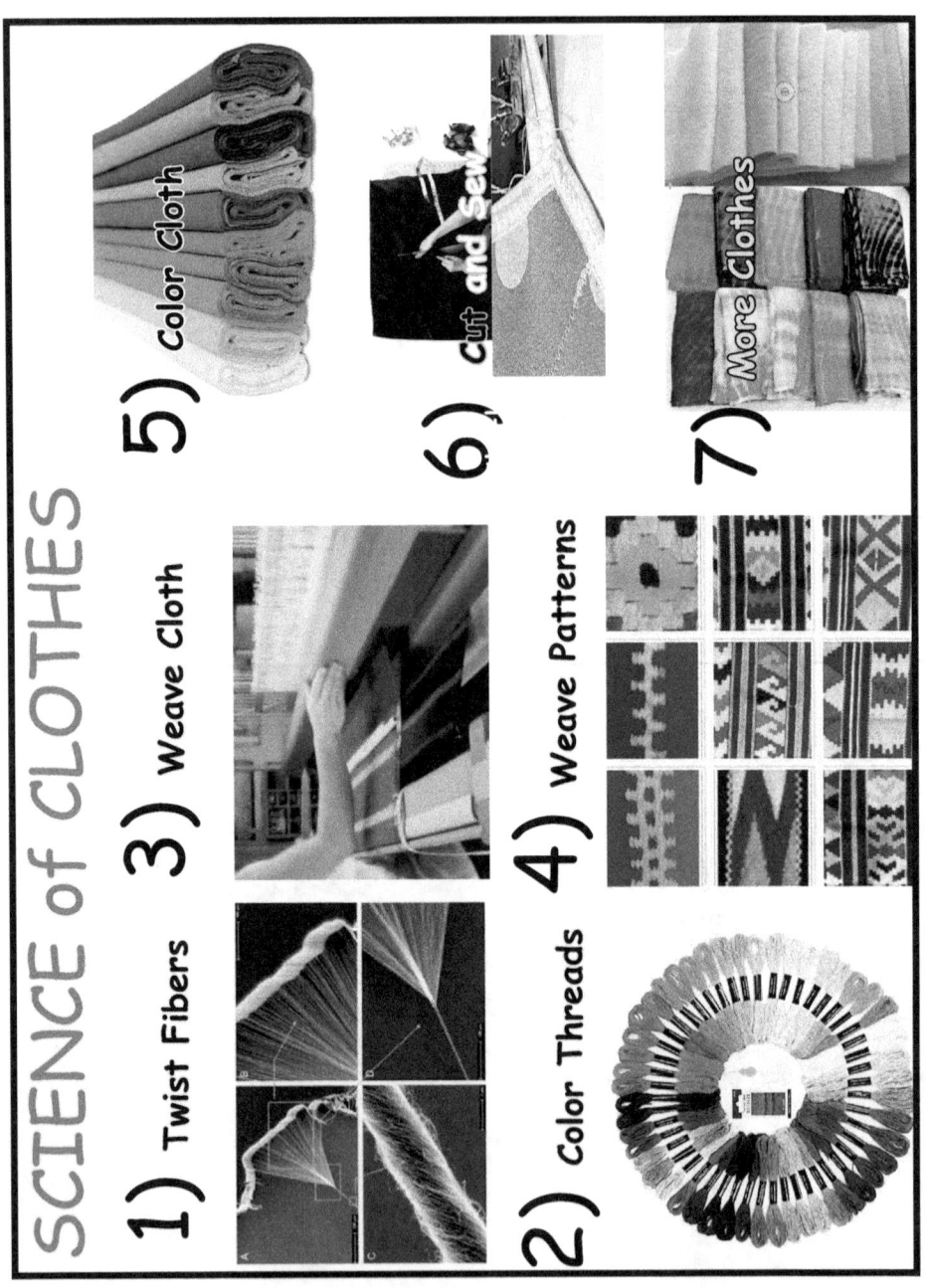

1) Twist Fibers

2) Color Threads

3) Weave Cloth

4) Weave Patterns

5) Color Cloth

6) Cut and Sew

7) More Clothes

Welcome to the
Science of Clothes!
Where do colored clothes
come-from?
Let's see the seven steps!

Back Cover

Cozy Clozy

Cozy Clozy — From Fibers To Fabrics
is a short story that shows the following. Why
are fibers made into things? How are clothes
made and colored? What things are made out
of fibers? Who uses fabrics? The brief, fun
flowing words complemented with colorful and
clear pictures share a story children love to
hear. After reading the book, children take
an active interest in cloth. They notice
different fibers and fabrics in their world.
Adults also find themselves looking at
labels and understanding textiles more.

37

Cozy Clozy

What Is It?

The STEM-Zen Program

is an integrated SCIENCE Program with thousands of pages and over 50 videos. Teachers help students go from science empty to knowledge enLighted!

STEM-Zen Program Strategy

Integrated Science Curriculum

STEM-Zen
Program Guide

STEM-Zen PREP
1) Cookie Come Froms
2) Cozy Clozy
3) Good Food Goes Bad
4) No Plants No Food
5) Plants Give
6) Sand Sea
7) Senses
8) Sun Above Clouds
9) Tad's Tale
10) Too Much Tech
11) Tree Trips
12) Wing Ways

Bonus
.Cats & Dinos
.Desert Rain
. Math
  — Numbers, Money, Shapes
. Home Stars
. Teams

STEM-Zen ONE
1) Seven Ideas
2) AIRPLANES
3) CARS
4) COMPUTERS
5) Smartphone-7 Waves
6) Smartphone
   — Objects Before Apps
7) Electricity
8) Everyday Objects
9) Stress Less
10) VIDEO GAMES

STEM-Zen ONE
Teachers Guide

STEM-Zen TWO
1) Science Thinks
2) PLANES
   — Past & Present
3) WEDGE TOOLS
   —Axes to Airplanes
4) Manu-FACTORY
   — Clay to Cars
5) COMPUTERS
   — Then & Now
6) NETWORKS
   — Wires to WiFi
7) PANDEMICS
   — Causes & Cures
8) MOON RACE
   — Chase to Space
9) Science of
   Lucky Stars

STEM-Zen THREE
1) Air, Water & Food
2) BOTS
   — Automata to AI
3) POWER
   — Windows to Wheels
4) NATURE
   — Where Life Lives
5) SEASONS
   — Turn, Tilt & Orbit
6) Images in Action
   — Why Movies Move
7) LIGHT
   — Sun to Screens
8) Bright Reading
   — Baas to Books

Science
by Subject